PSILOCYBIN MUSHROOMS OF THE WORLD

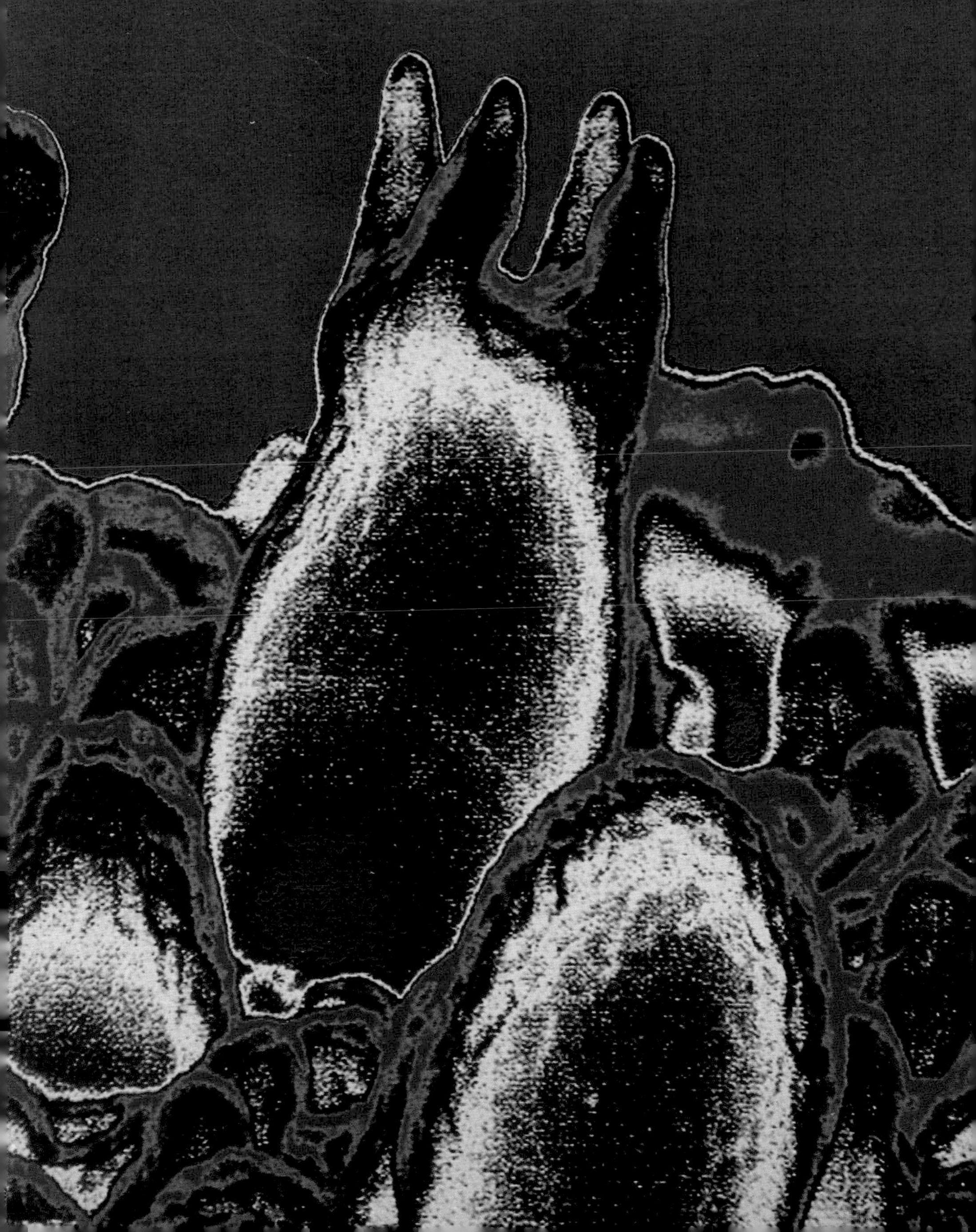

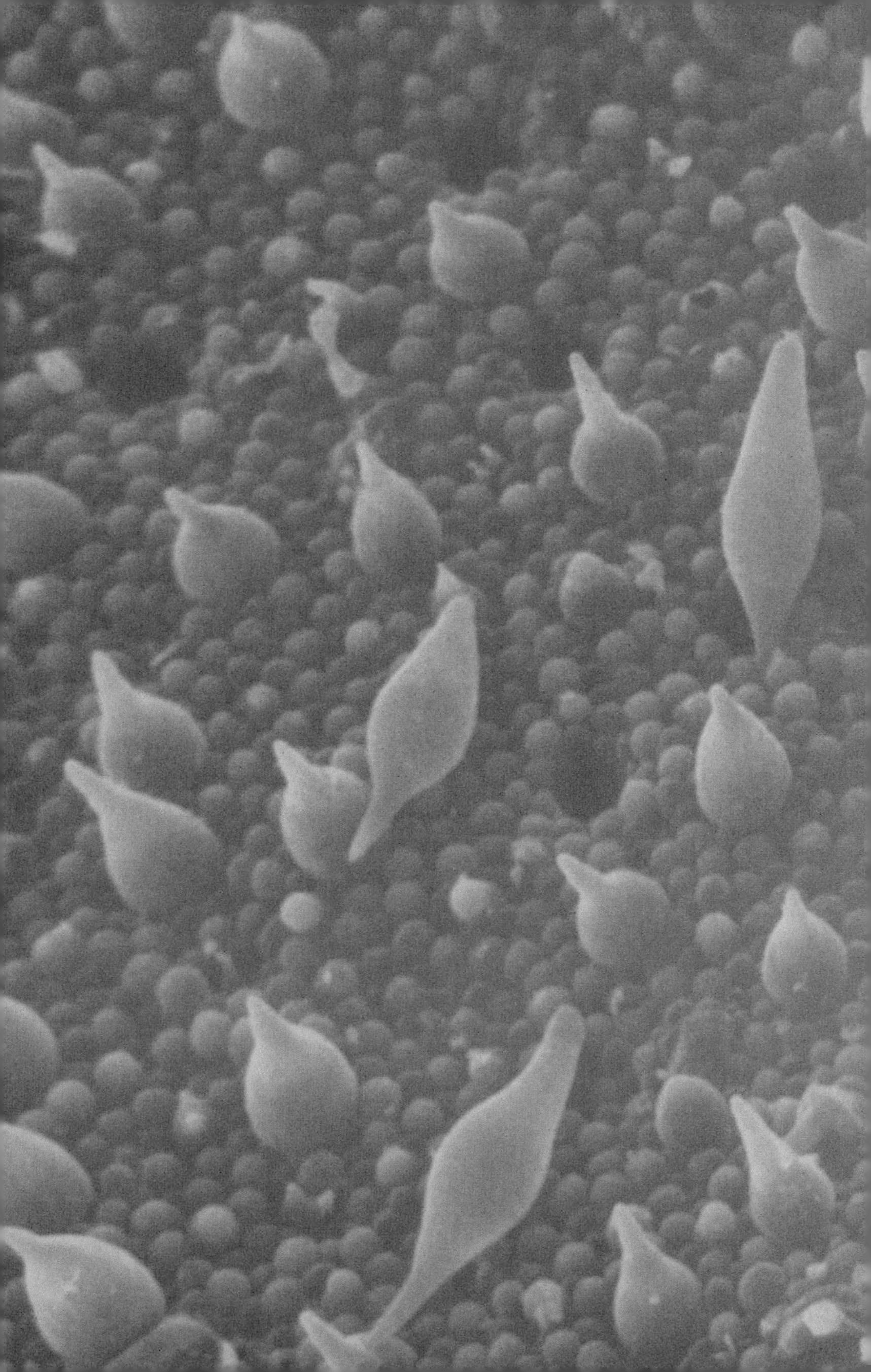

PSILOCYBIN MUSHROOMS *of the* WORLD

An Identification Guide

PAUL STAMETS

With a Foreword by
Andrew Weil

TEN SPEED PRESS
Berkeley, California

 Published in the United States by Ten Speed Press, an imprint of the Crown Publishing Group, a division of Random House, Inc., New York.
www.crownpublishing.com
www.tenspeed.com

Ten Speed Press and the Ten Speed Press colophon are registered trademarks of Random House, Inc.

Library of Congress Cataloging-in-Publication Data:
Stamets, Paul.
Psilocybin mushrooms of the world/Paul Stamets.
p. cm.
Includes bibliographical references (p.) and index.
1. Psilocybe-Identification. 2. Mushrooms, Hallucinogenic-Identification. I. Title.
QK629.S77S735 1996
589.2'22-dc20 96-15717
CIP

ISBN-13: 978-0-89815-839-7

Printed in China

Design by Catherine Jacobes

Photography and illustrations by John W. Allen: 73, 83, 140; Anonymous: 28, 59, 161, 166; David Arora: 70, 115, 123, 154, 171 (second from bottom); Harley Barnhart: 85; Michael Beug: 81 (right), 138, 151; Alan Bessette: 88; Arleen Bessette: 181; Jeremy Bigwood: 74 (left); Catherine Scates-Barnhart: 77, 81 (left), 105, 106, 132, 143 (top), 147, 157, 179, 184, 191, 194 (second from bottom), 197; Harley Barnhart: 85; Stan Czolowski: 197; Jochen Gartz: 99, 134, 182; Gaston Guzman: 92, 100; Kathleen Harrison: 17; James Q. Jacobs: 93, 103, 104, 116, 117, 118, 129, 130, 143 (bottom), 165, 168, 170; Grant Kalivoda: 12, 45, 246; Chris King: 90, 91; Paul Kroeger: 137, 171 (top), 194 (bottom), 197; Gary Lincoff: 110, 125, 171 (bottom); Jonathan Meader: back cover; Steve Morgan: 94, 141, 143 (middle), 158; Meinhard Moser: 145; Gyorgy Ola'h: 139; Steve Rooke: x; Giorgio Samorini & G. Camilla: 14, 23; Steven Schnoor: 200, 201, 202; Kazumasa Yokoyama: 87, 153, 155, 163, 183.

25

First Edition

Psilocybin Mushrooms of the World

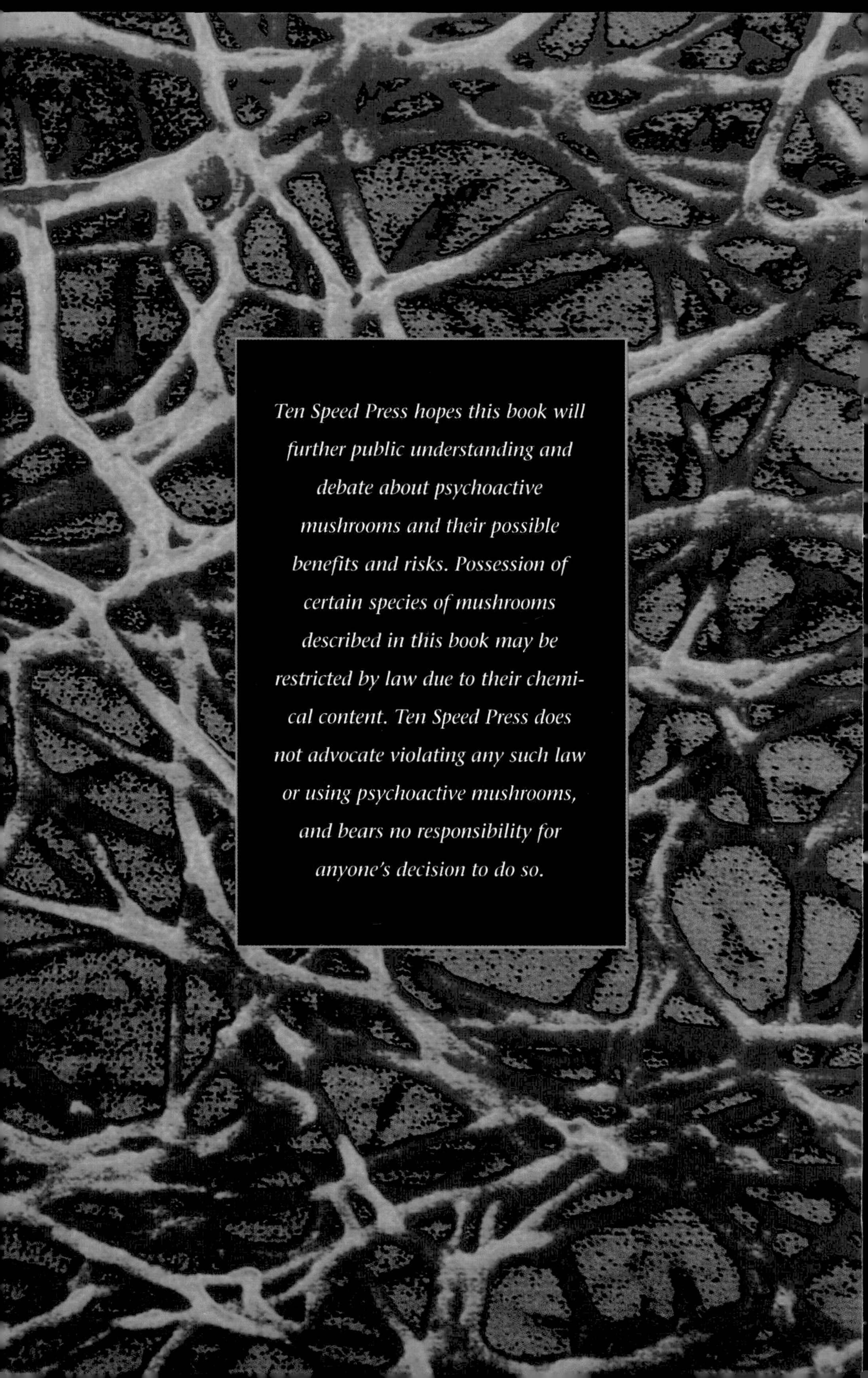

Ten Speed Press hopes this book will further public understanding and debate about psychoactive mushrooms and their possible benefits and risks. Possession of certain species of mushrooms described in this book may be restricted by law due to their chemical content. Ten Speed Press does not advocate violating any such law or using psychoactive mushrooms, and bears no responsibility for anyone's decision to do so.

Contents

Foreword

THIS GUIDE IS A UNIQUE ADDITION to the literature on mushrooms in general and psychoactive mushrooms in particular. Paul Stamets has brought together a mass of accurate information, both textual and graphic, on mushrooms that contain psilocybin—far more species throughout the world than have ever been presented in this format, including several species new to science. Anyone interested in these distinctive products of nature will find this book an invaluable reference, whether from the point of view of the collector, the scholar, or the prospective user.

Most psilocybin mushrooms are small, dull-colored fungi that never attracted much notice in our culture until R. Gordon Wasson brought the traditional ceremonial use of magic mushrooms in Mexico to world attention in 1957. For some time thereafter, large numbers of Americans and Europeans streamed to remote areas of Oaxaca in search of them, unaware that equally potent species grew in their own countries, sometimes literally in their own backyards. This guide makes clear that psilocybin mushrooms are ubiquitous, and, as more people search for them, they will probably be found in almost every place on earth.

Psilocybin resembles melatonin, serotonin, and other neuroregulators in its chemical structure. Its effects on human consciousness are profound. What is it doing in so many mushrooms? Certainly, today the presence of this compound is an evolutionary advantage because many humans find it attractive, thereby helping to propagate those species containing it. But what about in the past? Perhaps all that one can say is that psilocybin mushrooms are an illustration of the interconnectedness of all life and consciousness in ways that are more wonderful and strange than our intellects can explain.

Andrew Weil

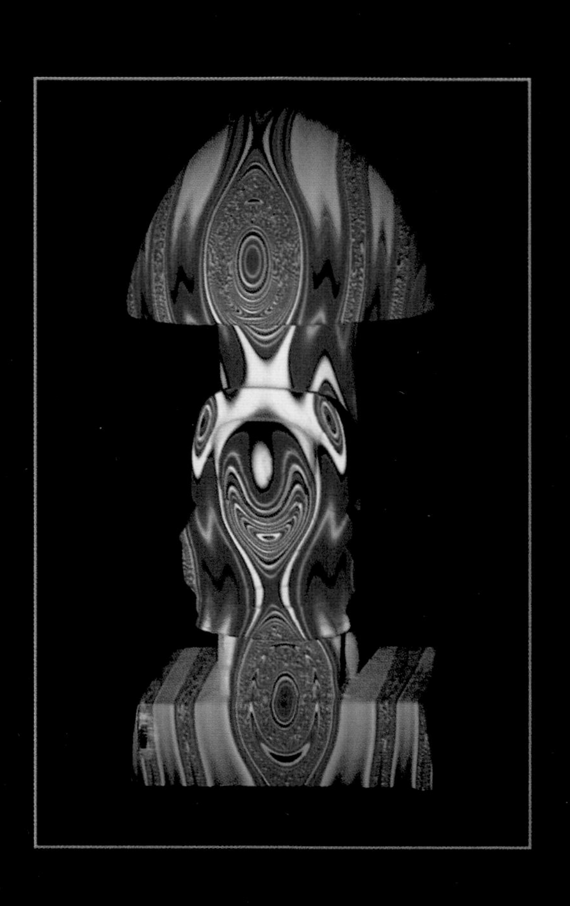

Introduction

I WAS NINETEEN YEARS OLD WHEN I embarked on my first book, *Psilocybe Mushrooms & Their Allies*. I was living in a mountain cabin near Darrington, Washington, and progress was slow and frustrating, in part because I was pounding away on a vintage Underwood typewriter whose keys required perpetual cleaning with toothbrushes. And yet, the project became a window into another dimension. Twenty years later, I am still collecting photographs and data on the subject. This book is an accumulation of research, both my own and my colleagues, through generations of experiences.

When my family first moved from a small town in Ohio to Seattle, I was mesmerized by my new horizons, marked by jagged, snow-capped mountains—a stark contrast to the bland scenery of the Midwest. On weekends, I would hike the trails of the North Cascades. I loved the rainforest—its smell, stillness, and sense of life quietly emerging all about me. Hiking up basalt-slotted canyons, fording over thundering waterfalls, or traversing ravines that led deep into the heart of dormant volcanoes, I found mushrooms everywhere. They lined the trails, bordered high alpine lakes, and dotted pristine meadows. Their shapes, sizes, and colors boggled the imagination, demanding recognition. An awakening began within me, one undoubtedly repeated for millennia. Mushrooms symbolized the bridge between life and death, between myself and the woodlands in which I lived. I sensed they were guardians of the sacred forests—conscious and watchful of my presence. Mushroom spirits soon enveloped my daily life, convincing me that they could be vehicles for greater good. I felt I had found my place within a continuum, and shared a sacred bond that spanned from the first paleolithic mycologists through the present, and to generations yet to come.

Against this backdrop of natural wonder, I worked as a logger, setting chokers for a living. The pay was good, and the dangers satisfied a primal instinct. My long hair set me apart from the die-hard, tobacco-chewing men who tried to outdo each other with demonstrations of bravery and stupidity. The job kept me in the woods, strengthened my resolve for self-reliance, was aerobically unbeatable, and provided an escape from the only other source of employment: dusty lumber mills, which I detested. During this tenure as a logger, I was introduced to the startling array of mushrooms that many of my logger friends collected and ate. Their excitement upon discovering each new mushroom patch was infectious. Soon, my mind become a sponge for information about fungi. The subject seemed to have no boundaries—the more I studied, the more I realized how little I knew. My path had been set on a course that continues to this day.

That the psilocybin-containing varieties would be absent from such a mycological paradise seemed impossible. I began searching in university libraries for information on the psilocybin mushrooms, but soon learned that very few books even had *Psilocybe* listed in their indexes. Those that did had one peculiarity in common: all descriptions, photographs, and otherwise useful information had been torn out, leaving a gaping and depressing testimonial to the eagerness with which others had sought the same information. At this time, the mid seventies, most people seeking Psilocybes would make long treks to Mexico. Few realized that these "magic mushrooms" were commonplace in regions of North America and Europe. The few books available, I soon discovered, had misleading if not outright false information. Many of the authors lacked field experience and simply copied the mistakes of their predecessors.

During the course of my research, I was surprised to discover that active Psilocybes are rarely found in the woods of the Pacific Northwest. Curiously, the potent Psilocybes are scarce in the wild but prolific and secure in their niche in the cities. The woodland *Psilocybe, P. pelliculosa,* is the one exception—it thrives in wild but disturbed grounds such as trails, abandoned forest roads, and other similar habitats. In twenty years, I have found only one specimen of *P. pelliculosa* deep within a natural forest. I am continually amazed that the majority of wood-decomposing Psilocybes thrive not in the depths of the wilds but in the disturbed habitats of densely populated areas, such as landscaping around buildings. As the use of decorative wood chips for landscaping became more common, a certain little brown mushroom began appearing with increasing frequency—a phenomenon that caught many off guard. Unfortunately, some of the world's most poisonous mushrooms also thrive in this habitat. Distinguishing between the groups is not difficult, but a simple mistake can have deadly ramifications.

While researching *Psilocybe,* I became accustomed to meeting great resistance from professional mycologists, many of whom had an instant distrust of anyone expressing a passion for *Psilocybe.* There were some mycologists who stated publicly that it would be better for people to die from mistakes in identification than to provide them with the tools for recognizing a *Psilocybe* mushroom. This bizarre attitude towards *Psilocybe* mushrooms and the people who used them reflected a chasm between generations.

Some physicians even seemed to take a perverse pleasure in the needless pumping of stomachs of patients who had consumed psilocybin mushrooms. One doctor told me he does so to "teach them a lesson." Ill-informed doctors, intoxicated with the power of their presumed authority, gave themselves license to espouse anti-mushroom rhetoric that strayed far from the truth. Later, I discovered that the reactions of these doctors and mycologists were often simply a result of ignorance. Since the majority of the psilocybin mushrooms—unlike the common edibles—are rare in conifer forests, most mycologists seldom encountered them during their sojourns. These were the same mycologists whose expertise was relied upon by attending physicians.

Beginning in the mid seventies, a new subculture evolved from the fabric of the counterculture movement of the sixties. In the northwestern and southeastern United States, hunting for Psilocybes approached the status of a national sport. In certain pastures, dozens of mushroom hunters could be seen on a daily basis—stooping, squatting, slowly and methodically walking under the gaze of stupefied cows and sometimes hostile farmers. I once estimated that each day during the fall, several thousand people were hunting Psilocybes in the fields of western Washington. The wave of interest soon became an invasion—a pandemic and a cause célèbre for an entire generation.

Trespassing and illegal-possession cases clogged the courtrooms. Hospitals saw more accidental poisonings and overdoses than ever before. Law enforcement officials, weary of the onslaught, typically prosecuted violators for misdemeanor trespassing rather than felony possession. I attended one packed court hearing where thirty individuals, including a friend, all pleaded guilty to trespassing. Each paid a fifty dollar fine. The entire court proceeding was clouded with a circuslike atmosphere; to be prosecuted for mushroom picking was of course totally absurd.

Ironically, each one of those pickers—knowingly or not—became agents for dispersing spores into more and more habitats. To this day, the grounds around the county courthouse and sheriff's department remain one of my favorite places to find *Psilocybe cyanescens* and *Psilocybe stuntzii*. Other favored sites include college campuses, utility substations, hospitals, office complexes, and ornamental gardens.

By the mid eighties, whole cities were overrun with Psilocybes—from Vancouver, B.C., to San Francisco. The growth of suburbia was expanding the zones of colonization. In particular, the marketing of wood chips (beauty bark) for landscaping continues to drive the *Psilocybe* revolution. Guerrilla inoculations became commonplace. Legions of Johnny Appleseed types traveled throughout the land carrying cardboard boxes filled with white, ropy mushroom mycelium. Grateful Dead concerts became favorite sites for distributing *Psilocybe* cultures. Private patches proliferated, as well as "mushrooms-for-the-people" beds in public parks, arboretums, nurseries, zoos...virtually anyplace where sawdust was used. Public domain beds—planted or natural—attracted new enlistees and they, in turn, created satellite colonies. The result is a continually unfolding, exponential wave of mycelial mass. The yearly splitting and expanding of mushroom beds has created mycelial footprints from Washington to New York, from Arizona to Canada. Similar trends in Europe soon followed. Many of the people I've met tell me they are on a vision quest; they believe that the world will become a more spiritual and peaceful place with each new mushroom patch. Many feel a deep, ecologically awakened attachment to the Earth, and believe that they are crusaders saving the planet. At any rate, they are succeeding in expanding the domain of psilocybin mushrooms.

In the past twenty years, the once-rare Psilocybes of the Pacific Northwest have come to dominate the populations of mushrooms found in wood chips. Their prevalence is a testimony to evolutionary success. Psilocybes cannot be eliminated from urban habitats. I don't know of any means for getting rid of them that wouldn't result in an ecological catastrophe. Mushrooms will be present as long as there are plants. They are a direct index of a healthy and biodynamic ecosystem.

During the winter of 1975, my life was changed by an event that challenged my concepts of time and space, and called into question my notion of reality. A relative, after years of collecting mushrooms in Mexico and Colombia, was now attending school in Seattle and was on the lookout for Psilocybes. At that time, we knew of several yet-to-be-named species that were growing around the University of Washington campus. One mid-November night he telephoned me in great excitement. "I found them!" he shouted. I warned him that it is difficult and unwise to identify mushrooms over the phone, but I also knew that he is mycologically astute, having a good understanding of features critical for identification, so I said I would give it a try.

We went down the checklist. I asked what color the spore print was. Dark purple-brown to nearly black, he replied. Was there a bruising reaction? Yes, the veil and base of the stem turned bluish when damaged. Was the cap covered with a translucent skin that could be pulled off? Yes, he could peel it off by slowly breaking the cap apart. I was cautiously optimistic. From his description, it sounded as if he had found a *Psilocybe*—probably a new species soon to be identified in the literature as *P. stuntzii,* in honor of Dr. Daniel Stuntz, professor emeritus at the University of Washington. "How many did you find?" I asked. "You wouldn't believe it," he replied, teasingly. That was all I needed to hear. Early the next morning, I drove to Seattle. Sure enough, when I arrived at his house, beautiful clusters of *P. stuntzii* were laid out on the kitchen table. His roommates huddled nearby as I confirmed identification. "Where did you find them?" "Come with me," he responded, barely concealing his excitement.

We embarked down University Avenue, equipped with a dozen paper grocery sacks. Near a busy intersection, we parked our car beside a power substation, which was landscaped with wood chips. "We're here!" he announced. I looked around. There seemed to be more pavement than open ground. I was perplexed, not realizing what "here" meant. Then, out of the corner of my eye, I saw the patch. He had good reason to be excited.

The mother lode must have numbered into the tens of thousands. There were so many mushrooms that the caps tightly touched, creating an elevated plateau four inches off the ground. The force of the fruiting was so great that sticks, leaves, *everything,* was lifted up over a contiguous fifty-by-fifty-foot area....Since the patch was located directly across from a busy Seattle police station, we began filling the paper sacks in great haste. Several students, noting our distinctive posture, stopped to join in the foray. In a matter of minutes, all had been picked. To this day, I have never seen another outdoor patch with such a concentrated eruption of mushrooms.

Upon returning to the house, we were faced with the daunting task of separating, cleaning, and drying our collection of a lifetime. We started making mushroom smoothies: yogurt, bananas, and mushrooms, which formed a thick, mudlike concoction with a very unpleasant flavor. Since our research had already determined that *P. stuntzii* was not a potent *Psilocybe* compared to *P. cyanescens* or *P. baeocystis,* I gave the best advice I could. "Try thirty to fifty mushrooms," I suggested, with some bravado. My estimate was buoyed by the fact that these were all experienced trippers, all Yale graduates, good friends, and for the most part emotionally strong enough for a high dose—should I have slightly overestimated.

In twenty minutes we started to experience the first stages of liftoff. The first hour is often the most unsettling part of the experience; later stages bring a familiar reassurance. Two hours into the experience, we could sense a slowing of intensity, and at three hours we plateaued. The dose was strong and richly rewarding. Unfolding geometric patterns surged towards me in wave after wave of beauty and complexity. My thoughts centered on God, evolution, the living earth, the infinite universe, the forces of good and evil, the mystery of death, and the paradox of time. We carefully stepped through the thousands of Psilocybes surrounding us on newspapers, covering virtually every available square foot of floor space except for narrow walkways. The enormity of our discovery humbled us. Awestruck, we christened our find The Great Boat Street Patch.

An hour or two past midnight, about six hours from the first mushroom smoothie, I went to bed. Geometric patterns continued to light up my field of vision as I descended into sleep. Several hours later, in the twilight between sleep and wakefulness, a peculiar and strangely real dream enveloped me.

I was at college, desperately trying to return to my mountain cabin as if my life depended upon it. This sense of urgency preempted all other priorities. *Go back.* Go back quickly. In dream state, I drove hurriedly into the mountains. Then, turning a corner on a country road, I came into a broad river valley lit up with a cold, clear light. The valley had flooded. Floating, dead, and bloated in the frigid sunlight were hundreds and hundreds of cows. The dream abruptly ended and I awoke in a cold sweat, struck with a fear of impending disaster.

I went downstairs and told my friends. This was like no dream I had ever had; there was a particularly foreboding strangeness to it that struck to the very core of my being. I feared there would be something like a nuclear war...maybe the USSR would attack, the snow would melt from the heat of the nuclear fireballs, and cows would be killed from the ensuing floods! My friends, not taking me seriously, began to joke. However, one person was curious enough to ask when this catastrophe would strike. I told him it would happen soon....I did not know when, except that I knew it would be on a weekend. He pointed to a date on the calendar two weeks out, December 1, and I knew that was it. He wrote "Paul says Doomsday" on that date, and the conversation changed course.

Two weeks later, after torrential rains and nearly record-breaking snowfall in the Cascades, an unusual temperature inversion swept over western Washington. Temperatures soared in the mountains, and the sudden thaw turned brooks into raging rivers in a matter of hours. Trees, houses, and bridges were flooded. My cabin, located only twenty feet from a glacial creek, was in immediate jeopardy. I knew that if I could not return quickly, all could be lost—my reference books, my manuscripts, all my personal belongings. The next day, I drove back to Darrington only to meet closures at one bridge after another. Finally, I drove a circuitous route, adding nearly a hundred miles to my trek, to find that my cabin was still safe but now ten feet closer to the raging river. The next day, I packed everything and headed south to Olympia. As I entered the Snohomish Valley, I stared in disbelief at hundreds of cattle who, stranded by the rising waters, had drowned overnight. It was December 1, the exact day my dream had foretold. This single event shattered my concept of linear time. The future can be foreseen.

Now I knew what shamans have known for centuries: the psilocybin experience can facilitate precognition of the future—especially, as in my case, of an impending biological disaster. Now I understood why the Mazatecs and Aztecs affectionately referred to Psilocybes as divinatory mushrooms, genius mushrooms, and wondrous mushrooms. They recognized that mushrooms are powerful sacraments and a significant evolutionary advantage for those sensitive enough to heed the call.

This book will be your guide to these sacred mushrooms, giving you the necessary tools for safely identifying psilocybin mushrooms throughout the world. The path is ancient, noble, and for many, holy. I sincerely hope that you will discover the capacity of mushrooms to lead to a new type of consciousness. Be careful, observant, respectful, and wise. The mushroom will be your teacher.

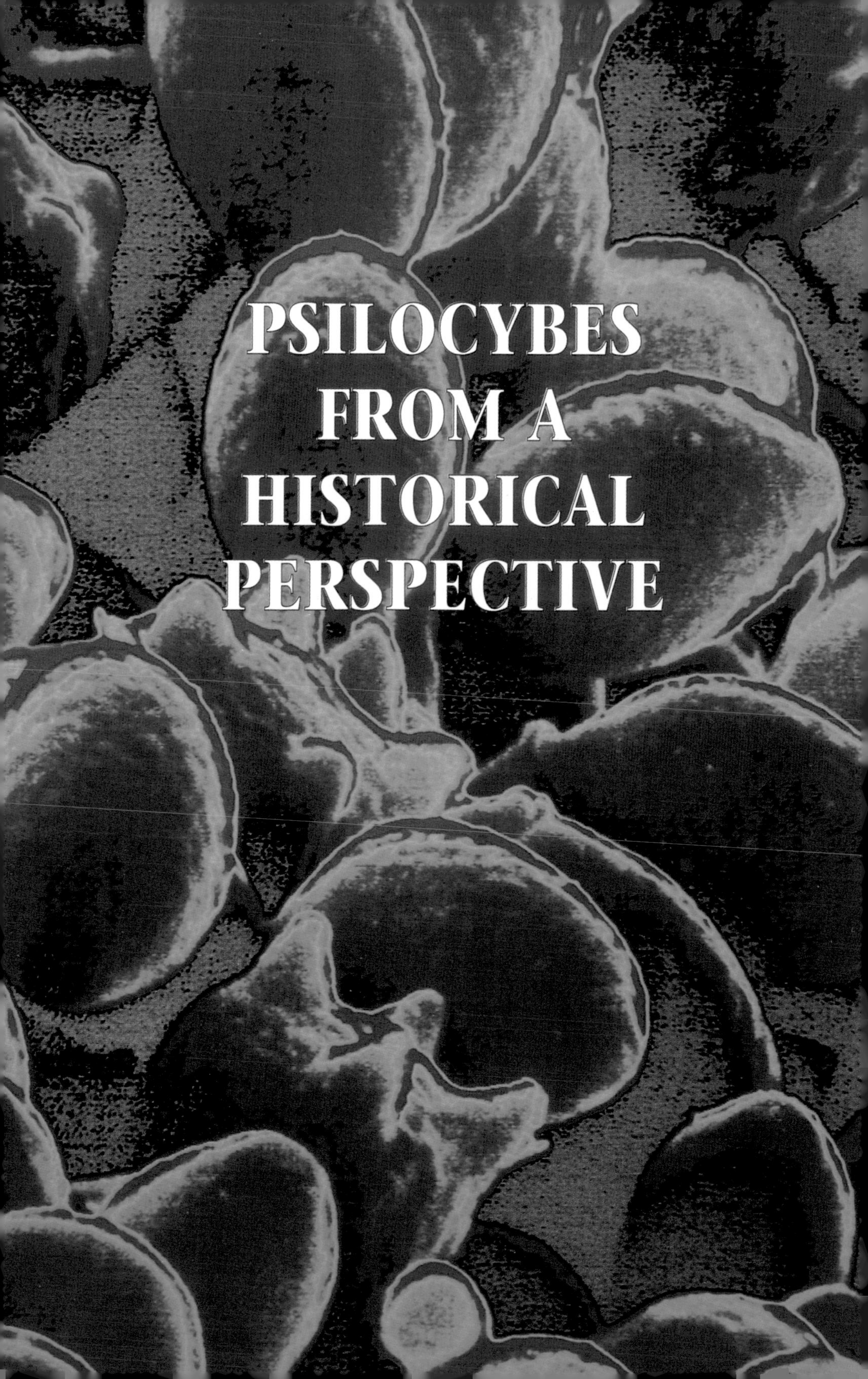
PSILOCYBES
FROM A
HISTORICAL
PERSPECTIVE

CHAPTER 1

Psilocybes from a Historical Perspective

THROUGH THE AGES, psychoactive plants have been used for religious and medicinal purposes. Mind-enhancing sacraments have been used by cultures widely separated by time and space, and have influenced many of the major religions and philosophies of the world. The sacramental use of mushrooms goes back at least seven thousand years, and probably extends to Paleolithic times.

Our understanding of the historical use of psilocybin mushrooms in other civilizations largely arises from the works of Blasius P. Reko, Richard E. Schultes, Roger Heim, and R. Gordon Wasson. First to re-discover and document the use of psilocybin mushrooms in shamanic ceremonies by indigenous Mesoamerican peoples, their research supported the theory that modern-day mushroom cults are the remnants of an ancient religion practiced by the Aztec and Mayan civilizations. These original ethnomycologists not only studied the ethnographic origins of mushroom rituals, they also had personal knowledge of their power.

The discovery of mushroom motifs and mushroom stones in excavations of Mayan temple ruins strongly underscores the important cultural role these mushrooms played.[1] One of the *Psilocybe* mushrooms (*P. mexicana*) was so esteemed as a holy sacrament as to be called *teonanacatl* (God's flesh) in the Aztec language. In the sixteenth century, a Franciscan friar, Bernardino de Sahagún, who travelled to the New World several decades after the expedition of the Spanish conqueror Hernando Cortés, reported the ritualistic use of teonanacatl by the Aztec peoples. However, misguided Catholic missionaries, in carrying out their campaign against "pagan idolatry," soon forced mushroom ceremonies into secrecy by persecuting those who were caught using them. As Christianity subjugated native rites, religions, and beliefs, artifacts—including mushroom stones and other motifs—were viewed by the conquering Catholics as idols to pagan gods and systematically destroyed.

A few of the estimated two hundred mushroom stones that escaped destruction, despite the concerted efforts of misdirected Catholic missionaries. The smaller mushroom stones are the ones found with metates, which were presumably used for grinding the sacraments prior to use (Borhegi 1961).

The widespread, *planned* destruction of this subculture nearly succeeded in completely erasing the ancient and rich cultural heritage of *Psilocybe* use in Mesoamerica. If it were not for the chronicling by Bernardino de Sahagún, who spent fifty years studying native cultures, we would have no eyewitness reports. Ironically, this priest from a foreign conquering nation remains our single best source of knowledge of *Psilocybe* mushroom use by the Aztecs—at least as reported through the eyes of his converted informants.

> At the very first, mushrooms had been served. They ate them at the time when, they said, the shell trumpets were blown. They ate no more food; they only drank chocolate during the night. And they ate the mushrooms with honey. When the mushrooms took effect on them, then they danced, then they wept. But some, while still in command of their senses, entered and sat there by the house on their seats; they danced no more, but only sat there nodding....And when the effects of the mushrooms had left them, they consulted among themselves and told one another what they had seen in vision.[2]

The Aztec emperor Moctezuma held an annual feast called the "feast of the revelations" in which the cognoscenti would eat green mushrooms. Unfortunately, the text detailing this feast has never been recovered. Wasson conjectured that the original text, reportedly voluminous, may have been suppressed by the church because of its controversial content. (Rituals involving mushroom use would have been considered satanic.) Nevertheless, despite the persecution wrought by the conquering Catholics, the ancient ceremonies persisted in secret, remaining hidden from outsiders until this century.

From the confrontation of two belief systems, a strange fusion of Christianity and mushroom ritual evolved. Maria Sabina, the renowned Oaxacan shaman who was "discovered" by R. Gordon Wasson's team, is a classic example. Her mushroom veladas were permeated with Catholic practices such as an altar to Christ, portraits of the Virgin Mary, and so on. When Wasson's "Seeking the Magic Mushroom" appeared in *Life* magazine on May 13th, 1957, millions of Americans were introduced to the *Psilocybe* mushrooms of Mexico. Roger Heim's excellent watercolors, accurate enough for field identification, reached into the heartland of America. Wasson's article introduced mushrooms to an entire generation of Americans during the formative, pre-sixties social period.

Soon, our fascination with psilocybin mushrooms could be measured by the thousands of aficionados who made the trek to Mexico in search of the famous Maria Sabina and her magic mushrooms. Her sudden popularity was unexpected and greatly regretted by Wasson. And yet, were it not for the passion of Wasson and his colleagues, these ritual remnants may have been lost forever. What is it about these mushrooms that has so enamored their followers?

The use of fungi as sacraments has historical precedence in European cultures as well. Aristotle, Plato, Homer, and Sophocles all participated in religious ceremonies in Greece

ABOVE: Ancient Aztec mushroom mandala. LEFT: The god *Xochipilli* (prince of flowers). Note circular mushroom mandalas portrayed at the base. The incurved margins of these mushrooms are taxonomically correct, illustrating that the artist had intimate knowledge of Psilocybes. These incurved margins are representative of forms of *Psilocybe caerulescens*, the landslide mushroom. Mexicans have chosen Xochipilli as the centerpiece of their 100 peso bill.

Persephone receiving from the goddess Demeter a mushroom central to the Eleusinian ceremonies, circa fourth century B.C.

where an unusual temple honored Demeter, the goddess of agriculture. For over two millennia, thousands of pilgrims journeyed fourteen miles from Athens to Eleusis, paying the equivalent of a month's wage for the privilege of attending the annual ceremony. The pilgrims were harassed by cajoling citizens on their journey to the temple.

Upon arriving at the temple, they gathered in the initiation hall, a great telestrion. Inside, pilgrims sat in rows on steps that descended to a hidden, central chamber from which a fungal concoction was served. An odd feature was an array of columns, beyond any apparent structural need, whose purpose remains a mystery. The pilgrims spent the night together and reportedly came away forever changed. In this pavilion crowded with pillars, ceremonies occurred, known by historians as the Eleusinian mysteries. Under the punishment of imprisonment or death, no revelation of the ceremony's secrets could be mentioned. These ceremonies continued until they were repressed in the early centuries of the Christian era.

In 1977, at a mushroom conference in western Washington, R. Gordon Wasson, Albert Hofmann, and Carl Ruck first postulated that the Eleusinian mysteries centered on the use of psychoactive fungi. Their papers were later published in a book entitled *The Road to Eleusis: Unveiling the Secret of the Mysteries*. That Aristotle and other founders of

Western philosophy undertook such intellectual adventures and that this secret ceremony persisted for nearly two thousand years underscores the profound impact that fungal rites have had on the evolution of Western consciousness.

If we are to consider Wasson and his colleagues the first generation of ethnomycologists, then Jonathan Ott, Terence McKenna, Andrew Weil, Christian Ratsch, Jochen Gartz, Giorgio Samorini, and other contemporaries could be considered the second generation. (See pages 218-229 for a listing of some of their works.) As the body of knowledge from this second generation amasses, a broad foundation is being laid in place for future ethnomycologists. The course of human history has been dramatically affected by the use of psilocybin mushrooms and will continue to be for years to come.

From the bronze door of a cathedral in Hildesheim, Germany, a bas-relief (circa A.D. 1020) depicting God, Adam, Eve, and the forbidden fruit—a taxonomic facsimilie of *Psilocybe semilanceata,* the Liberty Cap or Witches' Hat Mushroom.

---

[1] Although there is little doubt that the mushroom stones of Mesoamerica depict mushrooms, there is some debate about what mushroom stones represent. Borhegi (1961) found a cache of nine mushroom stones with nine metates, the mortar and pestle still used for grinding foodstuffs, including medicines. Guzman (1983) proposed that the mushroom stones actually represented edible species, such as *Boletus edulis,* because of their robust forms. However plausible, this hypothesis would not explain the recurring association with reptilian or jaguarlike base figures, animals intimately tied to the spirit world of cultures throughout Mesoamerica.

[2] From *The Florentine Codex,* by Bernardino de Sahagún (1985).

CHAPTER 2

Global Ecologies and the World Distribution of Psilocybin Mushrooms

PSILOCYBIN MUSHROOMS GROW throughout most of the world, and can be found in both fields and forests. Psilocybin mushrooms are saprophytes—they grow on dead plant material. Before the impact of human civilization, psilocybin species were largely restricted to narrowly defined ecosystems. Many thrive after ecological catastrophes. Landslides, floods, hurricanes, and volcanoes all create supportive habitats for many *Psilocybe* mushrooms. This peculiar affection for disturbed habitats enables them to travel, following streams of debris.

As humans destroy woodlands and engage in artificial construction, Psilocybes and other litter saprophytes proliferate, feeding on the surplus of wood chips and refuse, especially in the interface environments, wherever humans, forests, and grasslands struggle to coexist. Since human development seems inextricably associated with ecological disturbance, *Psilocybe* mushrooms and civilization continue to coevolve. Today, many Psilocybes are concentrated wherever people congregate—around parks, housing developments, schools, churches, golf courses, industrial complexes, nurseries, gardens, city parks, freeway rest areas, and government buildings— including county and state courthouses and jails! This successful adaptation is a comparatively recent phenomenon; in the not-too-distant past, these species were competing in a different environmental arena. Many of the Psilocybes are now evolving in a decidedly advantageous direction, parallel to human development. The way these mushrooms have evolved in close association with humans suggests an innate intelligence on the part of the mushrooms.

As human populations flourished, so too did the ranges of these litter-degrading mushrooms. Litter saprophytes thrive in the interfaces and relish the care humans provide through the devastation of landscapes. With the end of the last major ice age, about twelve thousand years ago, the melting glaciers riddled the exposed lands with rivers.

As climates shifted, new ecosystems appeared and continued to be transformed. Through millennia—either from natural or from man-made causes—many jungles evolved into savannas, and in many cases became deserts. Coincident with the retreat of the glaciers, the human species became less nomadic and more dependent upon planted crops. Many believe this marked the beginning of the path leading to civilization as we know it today.

Southern Algeria is one example. Today, the region is in stark contrast to its water-rich past. Once filled with rivers and lined with riparian woodlands, the Tassili plateau has now been engulfed by the expanding Sahara desert. In fact, the Tassili-n-Ajjer region was known as the "plateau of the rivers." In the 1930s and 1940s, hundreds of Paleolithic drawings were discovered in this region, painted on the walls of caves and rock faces. Ethno-archaeologist Henri Lhote and photographer Kazuyoshi Nomachi were the first to systematically catalogue the thousands of cave-art drawings. While searching for water, they accidentally encountered "a figure wearing a mask in a deep recess that may have been a sanctuary." The original artist lived seven thousand years ago, at a time when glaciers were rapidly receding. As these glaciers melted, estuaries etched through the flood plains. The glacial waters fueled the life cycles of many mushroom species. Time has erased much of the original detail, which showed many mushrooms outlining the shamanic figure. Fortunately, early photographs clearly communicate the intent of the artist—mushrooms were revered in a magico-spiritual context. The artist's intention is unambiguous: mushrooms were a powerful influence on his or her vision of the world. The beelike face may relate to the preserving of mushrooms in honey. For the Paleolithic human, the effects from ingesting psilocybin mushrooms would have precipitated one of the most phenomenal events ever experienced: a virtual cascade of consciousness, such as the awakening of the spiritual and intellectual self, the introduction to complex

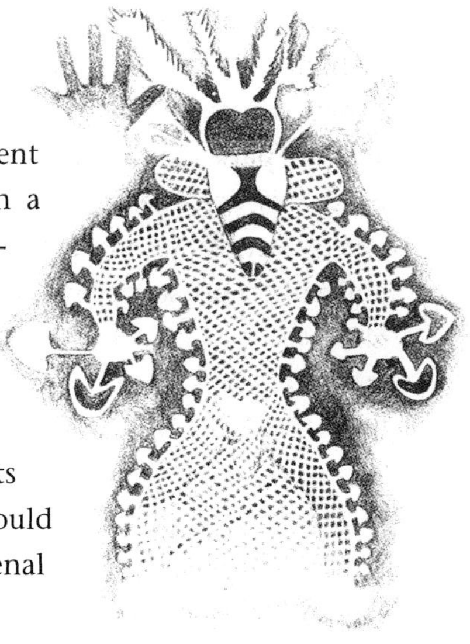

One of the oldest records suggesting mushroom use: cave art, approximately seven thousand years old, from the Tassili plateau in Southern Algeria. Redrawn from photographs by Lajoux (1961).

fractal mathematics, and the introduction to other dimensions. Such experiences continue to inspire artists, computer geniuses, and some of the greatest thinkers of this century.

One *Psilocybe* species is documented from northern Algeria: *P. mairei*, a *Psilocybe* resembling the potent *Psilocybe cyanescens*. This group thrives in riparian habitats—open areas with sandy soils seasonally littered with wood debris. *P. mairei* is relatively rare, having been collected only a few times this century. Do these few collections represent the end of a bygone era when mushrooms were more prevalent? Perhaps *P. mairei* is the same species that inspired the artist who drew the mushroom figures in the Tassili cave.

Other reports of presumably psilocybin varieties from northern Africa occasionally surface. Reports of a *tamu* (mushroom of knowledge) from the Ivory Coast are teasing but not sufficiently documented. The Italian researcher Giorgio Samorini (1995 b) noted that there are mushroom-based churches in southern Nigeria. Over the years, I have heard similar reports of Christian churches from Mexico, Brazil, and Russia featuring crosses whose centerpieces contained mysterious, encapsulated dried mushrooms of unknown identities and origins.

With the domestication of cattle, the dung-dwelling Psilocybes were brought within a defined geographical sphere of daily human experience. Pasture species such as *Psilocybe semilanceata*, the liberty cap, proliferated. Some researchers have suggested that *Psilocybe cubensis* (golden top of the old world) was imported into the Western Hemisphere with the Spanish missionaries and slave traders via the Brahman cattle they brought with them from islands off West Africa. *P. cubensis* soon became the most prominent dung mushroom throughout the tropics. Today, several hundred years later, *P. cubensis* can be collected from the dung of Brahman cattle in subtropical pastures circumnavigating the globe.

Nonnative mushrooms have also spread with the importation of exotic plants. Many species in the Pacific Northwest were undoubtedly brought from Europe, probably in the soil around the bases of exotic trees and ornamentals such as rhododendrons, roses, and azaleas. *Psilocybe cyanescens*, the wavy capped *Psilocybe*, is a good example. Every fall—when there are few visitors—I go searching for Psilocybes at rhododendron or rose gardens. Rarely am I disappointed.

Today, *P. cubensis* is the most commonly cultivated psilocybin mushroom in the world. Underground centers of cultivation, wherein large crops are grown, also function as invisible spore geysers, gushing germ plasm into immediate surroundings. Uplifted into the airstreams, spore clouds have spread across the continents. With the emission of so much spore mass, the range of distribution is likely to continue to expand. It seems that new strains could evolve, in our lifetime, with tolerances for cooler and/or drier environments. And with modern means of travel, spores can be carried thousands of miles in the course of a day—they can simply hitchhike upon unknowing airline passengers. I know of some people who have publicly opposed the spread of information about *Psilocybe* but have unwittingly spread spores through casual contact with it.

Psilocybes have propelled themselves to the front lines of the evolutionary race precisely because of their psilocybin content. The production of psilocybin has proven to be a competitive evolutionary advantage. Psilocybin mushrooms carry with them a message from nature about the health of the planet. At a time of planetary crisis brought on by human abuse, the earth calls out through these mushrooms—sacraments that lead directly to a deeper ecological consciousness, and motivate people to take action.

Throughout the world, at least thirty thousand mushroom species have been documented. About a hundred are known or suspected active species and varieties. By active, I mean they produce psilocybin, psilocin, baeocystin, or nor-baeocystin. The species producing psilocybin analogs are clearly concentrated in the genus *Psilocybe*, which has more than eighty species (Stijve 1995). A few psilocybin mushrooms belong to other genera, including *Panaeolus*, *Pluteus*, *Gymnopilus*, *Conocybe*, and *Inocybe*. Although the vast majority of the species in these genera are not active, more than half of the species in the genus *Psilocybe* are psilocybin producing.

Psilocybin mushrooms from the genera *Psilocybe* and *Panaeolus* are fairly safe to identify, in that there are no known poisonous species in those two genera. There are, however, several nasty species in the genera *Conocybe* and *Inocybe* that could be damaging or could kill you. Because of the danger of misidentification, I recommend that you avoid the genera *Conocybe* and *Inocybe* until you become sufficiently skilled at identification.

I spent hundreds of hours hunting in woods and fields before finding my first *Psilocybe*. I did run across many small, brown mushrooms and hoped they might contain psilocybin, but subsequently learned that the ones I had collected were poisonous! Today, I am grateful that my eagerness in finding these mushrooms was tempered by a prevailing concern for self-preservation. Knowing that many people are not as cautious convinced me that a good guide was urgently needed.

In the Pacific Northwest, at least four thousand mushroom species have been identified, with more than a dozen of these containing psilocybin. In Europe, about three quarters as many have been reported thus far. Mexico is the richest in psilocybin mycoflora. In fact, I have yet to find a single temperate or tropical habitat, with high annual rainfall, that lacks psilocybin mushrooms. But without some form of guidance, the random discovery of a psilocybin mushroom is, frankly, remote. In any region of the world, psilocybin mushrooms are greatly outnumbered by toxic mushrooms.

In some parts of the world, psilocybin mushrooms have not been reported at all. But just because they have not been reported does not mean they do not exist. Perhaps the indigenous population is simply unaware of them. Or, perhaps those who are knowledgeable are reluctant to discuss the subject.

Until recently, no psilocybin mushrooms had been reported from the woodlands of Colorado, despite more than fifty years of concerted efforts by competent mycologists. Many mycologists concluded that psilocybin mushrooms simply did not exist there. Then, in 1993, a lone specimen of a strongly bluing species of *Psilocybe*, otherwise resembling *P. pelliculosa*, was found in the high-alpine wilderness above Telluride, Colorado. (At approximately 10,000–11,000 feet.) The rarity of this find cannot be overstated. Was this a chance discovery of a species on the verge of extinction? We know now that it is a new species. Perhaps, some would say, this mushroom has remained hidden only to call out to the chosen one who found it, so she could give it to me to clone. Now that cultures have been established, its future seems assured. From an evolutionary point of view, its psilocybin content has directly guaranteed its survival. One must wonder if the production of psilocybin is being selected out as a beneficial evolutionary trait for this and other Psilocybes.

With the tools outlined in this book, anyone should be able to determine whether or not a mushroom is a psilocybin-containing variety. Of course, identification can be difficult and dangerous for the poorly prepared. However, armed with accurate information and a cautious approach to collecting, hunting for psilocybin mushrooms can be safe and fun. This book will give you a systematic approach, identifying key features that will allow you to quickly narrow the field of potential candidates. Armed with such knowledge, success on your forays will be made much more likely.

A poporo (70% gold; 30% copper) from the Quimbaya culture of Andean Colombia, circa 900–1200 A.D.

CHAPTER 3

Targeting Six Classic Habitats

IN MANY CASES, THE COLLECTOR CAN zero in on a species by choosing the correct habitat. For instance, the temperate liberty cap (*Psilocybe semilanceata*) is easy to find by searching through swampy sheep and cow pastures in the fall. In North America, this species ranges from Northern California to British Columbia. *P. semilanceata* grows in England, Ireland, France, Germany, Holland, Belgium, Sweden, Norway, Switzerland, Italy, Chile, South Africa—in other words, throughout much of the world. This species has even been reported from New Zealand, and probably grows unreported in other regions of the world. What ecological niche first gave rise to this species, and how long ago?

In long-established treeless pastures, the likelihood of encountering a deadly poisonous mushroom resembling a *Psilocybe* is fairly remote. However, new pastures—created by cutting back a forest—complicate the general rules about habitats and mushrooms. Habitats in transition, typically from forestlands to grasslands, will phase in diverse mushroom populations. In places like the Pacific Northwest, the importance of habitat has to be de-emphasized as a key feature. These complicated habitats—actually one habitat mixed into another—undergo radical transformations and consequently can easily confuse the untrained observer. Habitat as a target indicator is a far more useful feature in those environments that have achieved ecological autonomy and stability than those that are in transition.

Since all the Psilocybes and other psilocybin-producing mushrooms are saprophytes, they can successfully exploit a broad range of ecological niches and hence are geographically widely dispersed. As forests are cut and grasslands expand, a new mix of mushroom species surges in response. During this transitional period, the woodland and grassland species can often be found together. Housing developments created from cutting back forests are built upon soils rich with wood debris, making this perfectly suited for the wood-loving Psilocybes. When lawns are installed on top of this wood-enriched habitat, lignicolous and grassland species co-occur in the same habitat. The care humans give to these yards, many with automatic sprinkler systems, are ideal for encouraging mushroom growth.

This is particularly true with *Psilocybe stuntzii*, a wood-decomposing mushroom from Washington and Oregon. This blue-veiled *Psilocybe* frequents football, soccer, and baseball fields, and the landscaped areas around schools. When thousands of students dusted with spore mass leave these sites, they emit spore trails. This is but one method of dispersal that has proven especially effective in expanding the geographical range of *Psilocybe*. Upon germination, veins of mycelium thread paths into new habitats.

Some species of psilocybin mushrooms are truly habitat specific. As difficult as it is to find a truly natural habitat, I can pinpoint six distinct habitats in which Psilocybes and other psilocybin fungi flourish. These are classic habitats, but can often cross over into one another. Mushrooms atypical of the prevailing habitat can sometimes be found on island ecosystems. Or, after habitats have been disturbed, the mixed-in refuse can create debris fields wherein mosaics upon mosaics can overlay upon one another. The following six habitats are generalized, with some of the more prominent psilocybin and nonpsilocybin relatives highlighted.

An example of an island ecosystem: liberty caps, *Psilocybe semilanceata*, fruiting from a tuft of grass among inhospitable, gravelly soil. Italy.

Grasslands

Grassland habitats, especially wet swampy lowlands, support many of the tall, thin, small, conic-capped Psilocybes such as *P. strictipes, P. liniformans, P. semilanceata, P. mexicana,* and *P. samuiensis.* A nonpsilocybin-producing species, *P. inquilina,* fruits directly from the matted bases of field grasses, as does a recently discovered Indian species, *P. kashmeriensis,* that saprophytizes camel grass or lemon grass (=*Cymbopogon jawarancus*), an economically signficant resource for various oils, including citronella. *P. semilanceata* successfully exploits the dying rhizomes of many field grasses, including fescues (Keay and Brown 1990). By implication, tryptamine-producing grasses could have a potentiating effect on the production of psilocybin and/or psilocin (Gartz 1989). Canary *(Phalaris)* grasses, known for their high dimethytryptamine content, might make excellent companions for the coculture of Psilocybes.

Curiously, many of the grassland psilocybin-producing mushrooms also form sclerotia in culture. When dry, sclerotia look like hardened, nutlike structures. In a dormant stage, they can survive environmental disasters—a protective mechanism for surviving recurring fires that tend to sweep over grassland environments. Some of the grassland species known to produce sclerotia are *Psilocybe mexicana, P. semilanceata, P. tampanensis,* and *Conocybe cyanopus.* (For cultivation techniques, see Stamets and Chilton 1983 and Stamets 1995.) I know of no studies exploring the relationship between grasses, their tryptamine precursors, the potentiation of psilocybin and psilocin, and/or their relationships, if any, to sclerotia. This model is probable because Gartz (1989) was first to demonstrate that the raising of tryptamine levels in the substrate increased the psilocybin content in the harvested mushrooms.

Grassland habitats include many of the humus-loving species—those mushrooms thriving in a variety of soils, from red clays to dark loams. Those habitats pocked with islands of tall grass are usually easy to hunt in. The borders along forestlands are naturally cooler, are often the best places to pick, and have the longest fruitings, especially during drier weather. The types of grasses associated with Psilocybes are fescues, bent grasses, canary grasses, perennial ryes, sedges, and duneland grasses. For most species, grasslands grazed by sheep, horses, cattle, yaks, water buffalo, bison, llamas, or other domesticated animals tend to be more prolific.

Dung deposits

Dung is a great supporter of mushrooms. Since dung deposits are usually in grasslands, grass-loving Psilocybes can appear in the same geographical niche, especially as the dung disintegrates. Since dung deposits are short-lived habitats, the mushrooms that flourish do so in a matter of days. Co-occurrence of *Psilocybe mexicana,* a grassland species, with *Psilocybe cubensis,* a dung-dwelling species, is one example. The most prominent species to exploit the dung niche are *P. cubensis, Psilocybe coprophila, Panaeolus cyanescens (= Copelandia cyanescens),* and *Panaeolus subbalteatus.* Nonpsilocybin species, such as *P. coprophila* and *Psilocybe merdaria,* are common on dung along with many *Panaeolus.* A petite species, *Psilocybe angustispora,* favors marmot and elk dung in the Cascade mountains of the Pacific Northwest. Throughout the world, the noble dung heap provides a dependable and easy-to-find habitat for many psilocybin mushrooms as well as their close relatives.

Riparian zones, disturbed habitats, and gardens

These habitats appear suddenly—often cataclysmically. Unless the habitat is undergoing a major seasonal re-disturbance, the resident mushrooms will move on in a few years. I've identified three broad, interrelated subcategories that are distinct in themselves but can support many of the same mushrooms.

Riparian zones

These habitats are created from flooding rivers. Swollen rivers erode away soils, trees, and just about anything else in their path. The alluvial plains they create are characteristically high in sandy silt. An abundance of scattered, broken wood fragments characterizes this habitat. Tangled root balls of trees collect debris as the high waters recede. Deciduous trees such as cottonwoods (*Populus*), alders (*Alnus*), and/or willows (*Salix*) predominate, with assorted understories of grasses and brushes. These areas are fairly open, so are often sunny. Many of the species found in the riparian zones will also grow in disturbed habitats. *Psilocybe azurescens* and *Psilocybe quebecensis* are examples of psilocybin mushrooms found in riparian zones.

Geysers and hot springs are examples of a blending of riparian and disturbed habitats. One *Psilocybe*, similar to *Psilocybe cyanofibrillosa*, occurs in sedge grass at a hot springs near Whistler, British Columbia. One of the most geologically unstable regions of Canada, these hot springs continue to feed an alluvial plain. This microcosm environment, undergoing radical and continuous renewal, is a perfect springboard for riparian Psilocybes. Undoubtedly, hikers to the hot springs distribute spores as they return to their homes.

Disturbed habitats

Examples include roads punched into a forest, the grounds around a construction site, and landslides. Each of these habitats will evolve, and support fewer saprophytic mushrooms as the resources are consumed. Hence, the first two years tend to be the most prolific, with the third year declining. After four or five years, the habitats virtually expire. If new debris is introduced, however, or if the soils are upturned, these habitats can rebound with additional populations of Psilocybes. *Psilocybe caerulescens* is a classic example, and is even called *derrumbes* (landslide mushroom) by Mexicans because of its strong association with disturbed grounds.

I have also found Psilocybes in blackberry and Scotch broom thickets, where the environment is open, moist, and punctuated with deciduous trees. They are difficult habitats to explore, and they harbor rodent populations—whose burrows can be sheathed with mycelium. This phenomenon has been observed with *Psilocybe azurescens* near Astoria, Oregon. One knowledgeable collector reported to me that he has seen rodent pellets sprout *P. azurescens* mycelia. He speculated that the dung pellets serve the same function as the nutrient media in petri dishes used by cultivators, and that this is nature's way of encouraging concentrated spore germination.

Gardens

By accident or design, gardens are rich environments for producing psilocybin mushrooms. Because of tilling practices, gardens clearly fall into the disturbed-habitat category. Gardeners unwittingly cultivate mushrooms by importing exotic plants, amending soils, composting refuse, and faithfully watering. Furthermore, the inclination for gardeners to introduce the manure of cows, horses, or even "zoo-doo" also makes gardens hospitable for psilocybin mushrooms. Gardens rich in

A colony of *Psilocybe cyanofibrillosa* fruiting beneath a rhododendron. Water dripping from the leaves, combined with the shade, raises humidity and stimulates fruiting.

manure often sprout *Panaeolus subbalteatus* in temperate climates, and *Panaeolus cyanescens* in subtropical zones.

Vegetable gardens tend to support the terricolous Psilocybes, while ornamental gardens, richer in wood debris, support more lignicolous species. Rhododendron and rose gardens are often annually renewed by the activities of those caring for them: one rhododendron gardener I know places cardboard around the bases of rhododendrons every year to suppress weed growth. For twelve years straight, this garden has supported *Psilocybe cyanescens*. Other successful exploiters of this niche include *Psilocybe baeocystis, Psilocybe caerulescens, Psilocybe stuntzii*, and *Psilocybe subaeruginosa*.

Woodlands

This category is the most expansive of the six listed. Altitude, temperature, and rainfall broadly delimit the different sylvan ecosystems. There are temperate versus subtropical to tropical climates, each encompassing a diverse array of deciduous and coniferous trees. Psilocybin mushrooms are rarely found in the deeper forests of the Pacific Northwest, but occur with frequency in the pine forests of Mexico and elsewhere closer to the equatorial subtropics. Curiously, comparatively few psilocybin mushrooms grow in the woodlands of the true tropics—at least as far as we presently know.

Deciduous trees, especially cottonwoods (*Populus*), alders (*Alnus*), willows (*Salix*), box elders (*Acer*), sweet gums (*Liquidambar*), hornbeams (*Carpinus*), and similar trees thriving in moist soils along streams, ponds, and lakes support *Psilocybe*. In central Europe, *Psilocybe serbica* grows in deciduous forests interspersed with European beech (*Fagus sylvatica*), much as *Psilocybe caerulipes* does in the north central to northeastern United States in forests with American beech (*Fagus grandifolia*). However, Psilocybes are generally rare in these environments, unless the activities of collectors or cultivators result in localized population surges—events that, predictably, will occur with increasing frequency. (Techniques for cultivating *Psilocybe* mushrooms are covered in two of my other books, *The Mushroom Cultivator* and *Growing Gourmet & Medicinal Mushrooms*.) As described previously, deciduous woodlands devastated by recurring cataclysms tend to be better habitats for psilocybin fungi than those in stasis.

Coniferous forests, depending upon the region of the world, can be good habitats for Psilocybes. In the Pacific Northwest, I've found *P. baeocystis* growing directly from a Douglas fir cone (*Pseudotsuga menziesii*). Another sylvan species, *P. pelliculosa*, also shares a strong affinity for the needles of Douglas firs. I've seen *P. cyanescens* fruiting from Monterey pine (*Pinus radiata*). The collection and consumption of fir cones by squirrels results in piles of seeds at the bases of trees. In this rich duff I have found, although rarely, a *Psilocybe*. Gaston Guzman (1983) reports that the Mexican *Psilocybe muliercula* grows exclusively in *Abies* and *Pinus* forests at elevations of 3150–3500 and 2600–2800 meters, respectively. *P. caerulescens*, a lover of red soils and pines, was first reported near Montgomery, Alabama, in 1923. (It has not been re-collected since.) With more people searching for Psilocybes, the number of collection sites will expand, allowing for a better understanding of the ecology of these species.

• ***Psilocybe cyanescens* fruiting from a seed cone of *Pinus radiata*.**

Mosslands

Mosslands, such as those covered with *Sphagnum,* are generally poor habitats for psilocybin-producing species. The mossland habitat can exist as island ecologies within woodlands and/or grasslands. *Psilocybe atrobrunnea, Psilocybe montana,* and the former *Psilocybe corneipes* (Smith and Redhead placed this into a new genus, *Mythicomyces*) are representative examples. Few psilocybin-producing species are found exclusively in mosslands. However, I have found *Psilocybe baeocystis* and *Psilocybe cyanofibrillosa* growing directly from an island of moss in a rhododendron park. Related species of mushrooms that look very similar to the Psilocybes are the *Hypholomas (Naematolomas),* particularly the *H. elongatum, H. ericaeum, H. dispersum, H. polytrihca,* and *H. udum.*

Burned lands

Burned lands are not good places to find Psilocybes, but with time they can lead to good fruitings. In central Oregon, the practice of burning the grass-seed fields, and the subsequent regrowth, has given rise the subsequent fall to enormous fruitings of *Psilocybe strictipes.* Burned lands undergoing natural rebirth can support Psilocybes, but by themselves, without revegetation, do not produce well. Burned lands suffer from increased erosion, exposure, and cracking of the earth. But subsequent to disturbance and flooding, *Psilocybe* mushrooms can be stimulated into growth.

CHAPTER 4

The Dangers of Mistaken Identification

MISTAKES IN MUSHROOM IDENTIFICATION can be lethal. Since this book focuses on the *Psilocybe* mushrooms and their psilocybin relatives, I strongly encourage readers to use this information in conjunction with several good general field guides to mushrooms (see page 215). The more knowledgeable you become about *all* kinds of mushroom identification, the safer you will be.

If you have *any* uncertainty about a mushroom, do not eat it! We have learned which mushrooms are poisonous or nonpoisonous through the experiences of the unfortunate. However, since few of the little brown mushrooms have ever been sought after as food, little has been known about their biochemistry until recently. As more people sought the psilocybin varieties, poisonings from mistaken identifications increased. Physicians and mycologists joined forces and published case histories of the distressed victims. The extreme foolhardiness of randomly collecting little brown mushrooms becomes terrifyingly real for those whose unreserved enthusiasm overrode admonitions for adequate preparation.

On Whidbey Island, Washington, in 1981, three people picked and ingested deadly Galerinas, mistaking them for Psilocybes. Fearing prosecution for possession of psilocybin mushrooms, they delayed going to a hospital for two days, which is the most critical period for effective treatment. The girl, sixteen years old, died six days later (Beug and Bigwood 1982b). Her death is a tragedy that could have been avoided if she had had access to the information in this book.

Some of the mushrooms belonging to the genus *Galerina* that resemble Psilocybes and can be deadly are *G. autumnalis, G. marginata,* and *G. venenata.* These contain amatoxins (cyclopeptides) similar to those found in the lethal species of *Amanita,* such as *A. phalloides, A. bisporigera,* and *A. verna,* the "destroying angels." Although Amanitas do not resemble Psilocybes, the general form of Galerinas closely parallels that of the Psilocybes, differing to the unaided eye only in the overall color of the spores and to a lesser degree in the color of the caps. *Galerina* spores are rusty brown in deposit, while *Psilocybe* spores are purplish brown. However, when the caps are dried and/or frozen, the color can be very similar. Another deadly rusty brown–spored mushroom, *Pholiotina filaris* (=*Conocybe filaris*) also contains these amatoxins. Learn how to identify these mushrooms and avoid them. (See page 192.) In addition to *P. filaris* and the Galerinas, there are probably more species containing amatoxins than we presently realize.

I often find *Psilocybe* growing within inches of *G. autumnalis* and *P. filaris*. In fact, I look upon *G. autumnalis* as an indicator species—when this poisonous mushroom is present, I look in the surrounding habitat for the *Psilocybe* that are likely to be nearby. One dramatic example of this occurrence is illustrated below. Two colonies of mushrooms, one *G. autumnalis* and the other *Psilocybe stuntzii*, have overlapped! The stems of poisonous and psilocybin-producing varieties in this case were actually touching. I worry that amateur collectors would not be knowledgeable enough to separate the deadly species from the psilocybin one. In their eagerness, how many would overlook this deadly dance of species? Would you? This chance find was one of the primary reasons that I was compelled to write this book.

The purple-spored *Psilocybe stuntzii* (left), a psilocybin-active mushroom, and the rusty brown–spored *Galerina autumnalis*, a deadly poisonous mushroom, growing side by side.

Psilocybe pelliculosa **sharing the same habitat with deadly Galerinas.**

Beware of *Galerina* species and *Pholiotina filaris,* as accidental consumption can result in an agonizing death. Following is a description of the effects of eating *Amanita phalloides.* Since the deadly Amanitas, Galerinas, and Conocybes all produce cyclopeptide toxins, similar effects can be expected after eating the above-named species of *Galerina* and *Conocybe.*

> First symptoms come late—six to twenty-four hours (average ten to fourteen hours) after ingestion of the mushrooms. Sharp abdominal pains are followed by violent vomiting and a persistent choleralike diarrhea (often containing blood and mucus). These symptoms tend to subside and the patient appears to improve. In three to four days the patient's condition begins to worsen, with symptoms of liver and kidney failure leading to death in seven to ten days. Autopsy findings are: marked gastro-intestinal edema, hemorrhagic gastro-enteritis, lymphoid tissue and lymph node hyperplasia, fatty degeneration of the heart and liver similar to that seen in carbon tetrachloride poisoning, tubular necrosis of the kidneys, and swollen brain with multiple hemorrhages and degenerative nerve cell damage. Death is primarily from liver and kidney failure....[3]

This description should convince any hunter of psilocybin mushrooms that mushroom identification must be approached seriously, and with the utmost caution.

Other brown-spored genera that have an abundance of poisonous species are *Inocybe, Cortinarius, Hebeloma*, and *Entoloma*. By taking spore prints, paying attention to detail, and following the keys in this book, the possibility of ingesting poisonous mushrooms is minimized.

Several species in this book, such as the bluing species formerly belonging to the genus *Stropharia* (ex. *Stropharia aeruginosa* = *Psilocybe aeruginosa*), have not been fully tested. Bluing Inocybes, like *I. aeruginascens*, although psilocybin containing, should be treated with extreme caution, because their close relatives can be very poisonous. In general, members of the genera *Psilocybe* and *Panaeolus* seem quite safe—only psilocybin, psilocin, and related compounds have been detected thus far.

---

[3] "Mushroom Poisoning in the Pacific Northwest." Seattle: Puget Sound Mycological Society, 1972.

CHAPTER 5

Good Tips for Great Trips

THERE ARE NO GUARANTEES FOR A GREAT TRIP with psilocybin mushrooms, as the experience can be profoundly different for each individual. Not all experiences are positive. For some, it can be dangerous. The experience is influenced not only by the amount and potency of the species but also a host of cultural, environmental, and psychological factors. Those eating *Psilocybe* mushrooms for recreational purposes often have experiences far different from those seeking religious insights. Irresponsible use can result in disaster for the participants and those around them. I have observed that individuals who fear losing control are more often psychologically predisposed for negative encounters. Those suffering from illness may be physically or psychologically unprepared, and should avoid experimentation. Although psilocybin mushrooms are empowering, it is the person using them who must take ultimate responsibility for the nature of his or her experiences.

When mushroom use is ritualized—guided by veteran users who can help orchestrate the necessary variables with careful attention to set and setting—an extraordinarily gratifying cerebral adventure can unfold. The senses come alive, elevated to a level far above ordinary consciousness. Vision becomes more acute. Hearing is enhanced. Awareness of the fluid state of our bodies is brought to the forefront. Some participants believe Psilocybes simply amplify the qualities of the inner self. Some feel the mushrooms help them understand their place in the universe and their intimate connection to the planet. Many have deep religious revelations, or feel that they have been forever changed for the better.

I am certain there will be people who will be aghast at what I am suggesting. Since people are naturally afraid of the unknown—or what they have not yet experienced—I am not surprised or offended by those who feel threatened by the following guidelines. However, given that people have used, are using, and will use these mushrooms, why not make some suggestions to minimize bad experiences and maximize the positive? These suggestions are solely my own. I trip on mushrooms only once or twice a year, to re-calibrate my cerebral and spiritual compass. This feels right for me. For the vast majority, psilocybin mushrooms are used infrequently. The effects are often so profound as to take months before the interest in taking them rekindles. They certainly are not addictive, as they tend to be self-limiting by nature.

Dosage: how many mushrooms to ingest?

The amount to take is dependent upon many factors, most importantly the species and your individual sensitivity. For first timers, who may approach the alteration of their consciousness with some trepidation, a low dose is recommended. Terence McKenna, psilocybin psychonaut extraordinaire and author of *Food of the Gods, True Hallucinations,* and *The Archaic Revival,* boldly suggests, "When in doubt, double the dose." He believes that only when you are slain with the power of the mushroom does the message become clear. He is right in that the higher dose will get your attention, but not everyone can tolerate intense trips. The unprepared can suffer from panic attacks. Some are afraid they may do something that they would not ordinarily do. This fear of the loss of self-control becomes a central issue amongst all users as their dosage increases. Those who are willing to let go, and who do not fear their inner self, seem better prepared to tolerate higher doses. They flow with, not against, the tide of the experience. Psychological predisposition, combined with other factors such as set and setting, the species, and sensitivity, makes dosage guidelines especially difficult to pinpoint.

Readers should be forewarned about individual sensitivities. The doses I describe *should* hold true for most of us. However, I know of two mycologists who have had abnormal reactions. The first needs only 1–2 grams. He reacts as though he has had 3–4 times the dose, however. During a recent session, a hike for him was simply getting off the floor. At the other extreme is the woman who did not feel any effects whatsoever at dosage levels exceeding 5 grams. Well-read on the history of psilocybin mushroom use, she had looked forward with great anticipation to the experience and felt cheated that no effects were felt.

Adding to the difficulty of ascertaining precise dosages is research by

Beug and Bigwood (1982a), which found a fourfold difference in the psilocybin content from mushrooms grown on rye grain, and nearly a tenfold difference in specimens collected in the wild. Gartz (1989) found that raising tryptamine concentrations of cow dung and rice media by 25 millimolars directly affected the potency of *Psilocybe cubensis* mycelia, specifically in psilocin content: from .09% to 3.3% of dried mass. (Psilocybin content was actually depressed, but not nearly on the same order of magnitude.) This discovery may well explain why certain strains of *P. cubensis* are more potent than others when the diets of cattle are supplemented with nutrient-enriched feed.

Jochen Gartz (1989) also made the interesting observation in the culturing of *Psilocybe azurescens* and *P. cubensis* mycelium that the raising of malt sugars to more than 10% resulted in the complete suppression of psilocybin production. These two observations, one with tryptamines and the other with (carbon rich) sugars, underscore that the nutritional content of the substrate significantly affects potency. Additionally, Gartz (1992) found that younger specimens are generally more potent than mature ones, an observation many users have also made. These are but some of the variables in a constellation of factors that complicate consistency in the production of psilocybin and psilocin.

Although no studies have been published, I suspect that ultraviolet radiation from the sun markedly lessens the potency of psilocybin-containing mushrooms. At an ethnobotanical conference in Palenque, Mexico, I observed that sun-struck *Psilocybe cubensis* appear weak in comparison to those found in the shade. Clearly, UV destroys molecular bonds, which would readily explain why sun-struck collections are less potent. A number of participants at this conference, well versed in psilocybin experiences, believe that ingesting several grams of Syrian rue (*Peganum harmala*), an incense seed rich in MAO (monoxidase) inhibitors, potentiates the effects of otherwise weak psilocybin-containing mushrooms. The reports are anecdotal; no control studies have been published to date. A simple experiment would be to simply establish what a "below-threshold" dose would be from a uniformly dried and powdered mushroom collection. Then, a week later, to ingest 3 grams of Syrian rue and the same "below-threshold" amount of powdered mushrooms. If strong effects are felt, the Syrian rue would logically be the potentiator. This method has been used successfully with many dimethytryptamine-containing plants that would otherwise not be active through digestion

(Ott 1993). Readers should note that self-experimentation with monoxidase inhibitors could allow other compounds that would otherwise be detoxified in the gastrointestinal track to pass through unaltered.

Mycologists have received scattered reports of unusual sensitivities to both psilocybin and nonpsilocybin mushroom species. Recently, a young girl from British Columbia who nibbled a small fragment of chicken-of-the-woods (*Laetiporus sulphureus*) was catapulted into a hallucinogenic experience that persisted for several hours, with hallucinations of "lines and shapes of bright colors" (Appleton et al. 1988, 48). Experts are at a loss to explain this reaction. Over the years, I have heard of several other incidents where people have eaten a species commonly regarded as a culinary edible only to have classic psychoactive reactions. How can these unusual reactions be explained? Can certain unique pairings of a mushroom and a human result in a cascade of neurological events that others would not experience? I wonder if each of us has a unique fungal partner that, upon pairing, catalyzes an extraordinary sequence of neurochemical changes. Since we are all different, I suggest initially screening yourself for your individual sensitivity by starting with low doses of any mushroom. Keep in mind that a complex sequence of variables are at play, any one of which can have a major influence on the subjective psilocybin experience.

Doses can be targeted to some degree by dividing the potency of the mushrooms by your body weight. First, you must know the general range of psilocybin and/or psilocin in the species you are ingesting. (Consult the psilometric scales on pages 39-40.) Since body mass can influence the pharmacological impact of psilocybin, doses should be adjusted upwards or downwards accordingly, before taking into consideration personal sensitivities. For the average adult male weighing 176 lb (80 kg), a manageable dose would fall into the .25 mg per kg range, or about 20 mg. For most people, a high dose would be .5 mg of psilocybin/psilocin per kg of body weight, or 40 mg. An extreme dose—too high even for most veterans— would be 1 mg per kg or 80 mg of psilocybin.

For species such as *Psilocybe semilanceata* (liberty cap), which can average about 1% psilocybin/psilocin content, 1–2 grams of dried mushrooms is recommended for first timers. This is equivalent to only 10–20 mg of psilocybin, and will be, for most people, a dose with moderate effects. The symptoms will last, in most cases, no more than 4–5 hours. When Maria Sabina gave R. Gordon Wasson thirteen pairs of fresh

Psilocybe caerulescens to ingest, he probably received well in excess of 50 milligrams, a dose dangerous for those not under the direct guidance of a gifted shaman.

At low doses, hearing will be enhanced, but true visual or audio hallucinations are unlikely. Often, a childlike giddiness is felt. Colors will seem brighter. I often feel a tickling in my stomach. Breathing can seem more labored and focused. There can be an increased sensitivity to temperature. Music unfolds with amazing intricacy and beauty. Overall, I have an enhanced awareness of my liquid state of existence. Perception on all levels seems increased. Psilocybin mushroom participants commonly share in this prevailing sensitivity of the human body and its functions.

Once you have become a veteran and feel comfortable with a given dose, the amount of mushrooms can be increased in single dried-gram increments, again using *P. semilanceata* as the standard. Over time, each individual will come to understand his or her personal psilometric scale of sensitivity.

The following analyses are derived from research conducted over the past twenty years. Considerable variation in the range of psilocybin and psilocin has been found within each species. For the purposes here and with but one exception, I am listing the maximum concentrations detected in twelve *Psilocybe* species. Please consult the references for more complete information, as other indole alkaloids have been found in these mushrooms besides psilocybin and psilocin. Baeocystin, norbaeocystin, and/or aeruginacine are closely related to psilocin, and may be active (Gartz 1992). In general, these related indoles are present in lesser concentrations than psilocybin and psilocin. The actual potency of the mushrooms you collect is likely to be less rather than more potent than the table indicates.

The percentage figures are always based on dry weight. For instance, a 1 gram mushroom containing 1% psilocybin would have .01 grams or 10 milligrams psilocybin. The threshold dose, the amount where pharmacological effects can be first noticed, is 2–4 mg. Stronger but still moderate effects, which Jonathan Ott (1993) describes as "entheogenic," are inspired above 6 mg for psilocin and 10 mg for psilocybin. Since psilocybin is degraded into psilocin during digestion, you are only feeling the effects of psilocin, a dephosphorylated form of psilocybin.

The psilometric scale of comparative potency of selected *Psilocybe* mushrooms

SPECIES	% PSILOCYBIN	% PSILOCIN	% BAEOCYSTIN	REFERENCE
P. azurescens	1.78	.38	.35	Stamets and Gartz 1995
P. bohemica	1.34	.11	.02	Gartz and Muller 1989; Gartz (1994)
P. semilanceata	.98	.02	.36	Gartz 1994
P. baeocystis	.85	.59	.10	Repke et al. 1977; Beug and Bigwood 1982(b)
P. cyanescens	.85	.36	.03	Stijve and Kuyper 1985; Repke et al. 1977
P. tampanensis	.68	.32	n/a	Gartz 1994
P. cubensis	.63	.60	.025	Gartz 1994; Stijve and de Meijer 1993
P. weilii (nom. prov.)	.61	.27	.05	
P. hoogshagenii	.60	.10	n/a	Heim and Hofmann 1958
P. stuntzii	.36	.12	.02	Beug and Bigwood 1982 (b); Repke et al. 1977
P. cyanofibrillosa	.21	.04	n/a	Stamets et al. 1980
P. liniformans	.16	n/d	.005	Stijve and Kuyper 1985

Readers should note that, within any one species, there can be a tenfold or more range in psilocybin and psilocin production from one collection to the next! Also, this table excludes potencies that have been potentiated through the addition of tryptamine precursors to the substrate, as in the Gartz (1989) experiments. Not withstanding the obvious difficulties that have already been stated, this rating system defines categories based on the following ranges and maximum potencies, based on a dry-weight percentage of psilocybin + psilocin:

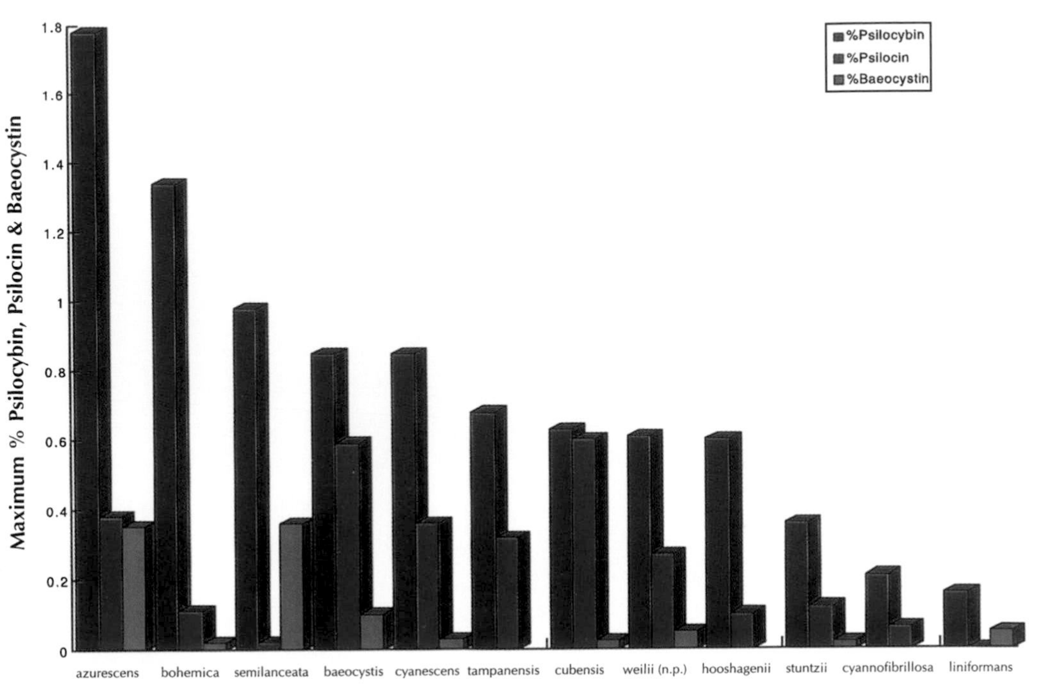

A comparison of maximum reported percentages (dry weight) of psilocybin, psilocin, and baeocystin in twelve species of *Psilocybe.*

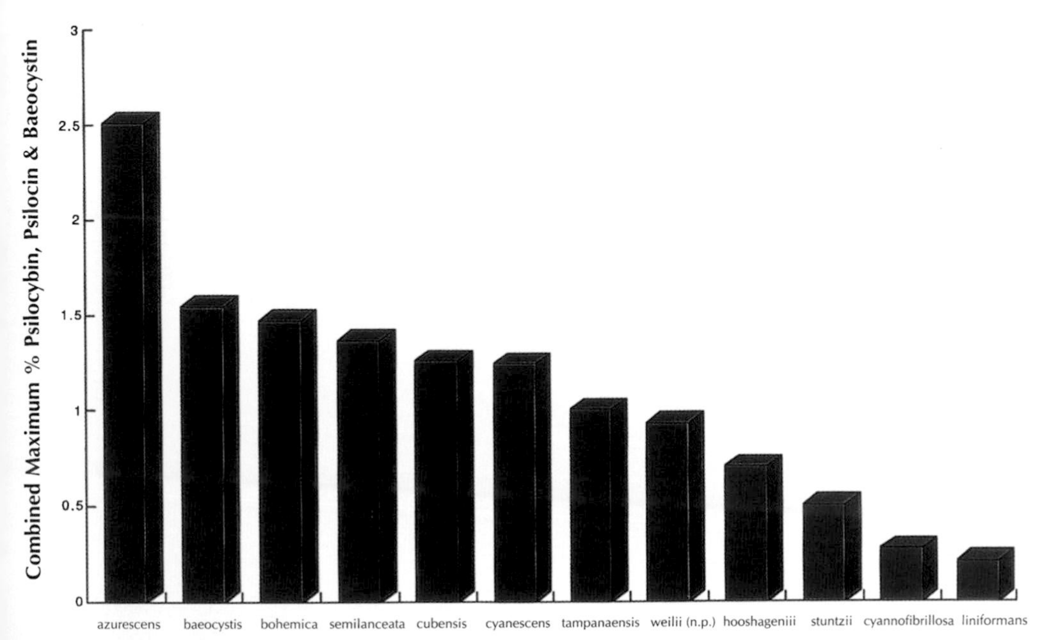

A comparison of the combined maximum percentages of psilocybin, psilocin, and baeocystin in twelve species of *Psilocybe.* The presumption here, not proven, is that all three are equally active and the effects are cumulative.

Potency rating scale

not active	0%
weakly active	<.25%
moderately active	.25–.75%
highly active or potent	>.75–2%
extremely potent	>2%

If studies prove that baeocystin is equally active to psilocybin and psilocin, the percentage of baeocystin would became a significant factor influencing overall potency.

With many of the species described in this book, chemical analyses have been reported in the scientific literature and have been, in turn, re-reported here. Please refer to the comments in each species descriptions, which will often refer you to bibliographic references.

Rituals: safety nets for the psychonaut

As the dosage increases, the need for ritual becomes increasingly important. Working within a ritual setting gives structure to the experience and progresses events along a positive path. Ritual can become a safeguard if the going gets rough—it can help lead you through the experience and make it profoundly meaningful. After repeated sessions, the ritual becomes a psychological road map, providing a framework for safe tripping. Rituals are built from the lessons learned from previous good experiences. But at some point, for the shamans amongst us, being safe is not the priority—pushing the envelope to new revelations is.

Before embarking on an adventure of this significance, you should be well equipped and prepared mentally and physically. Please consult "Twenty essential items for the sacred mushroom expedition," on page 49. Being prepared frees the experience from petty constraints and helps make it safe and worry free.

Again using *P. semilanceata* as a baseline species, 2 grams will bring on the first colorful geometric patterns, and lead to noticeable changes in auditory perception. At 3 grams and above, some people will experience visual waves. The wave phenomenon is especially interesting—it is as if the air were actually in a liquid state and tidal movements flow in from the distance, distorting the scenery as they project towards or away from you.

At 4 grams and above, the experience becomes more intense, lasting 4–5 hours. At 5 grams, the experience can be nearly overwhelming and span 6 hours, but rarely longer. Even for the most experienced, I do not recommend doses above 7 dried grams. At these higher levels, the set and setting must be especially supportive and safe. At higher doses, some people report a loss of muscular control and strength that can persist into the following day. I have heard this from individuals ingesting *Psilocybe azurescens, Psilocybe cyanescens,* and *Panaeolus subbalteatus*. No plausible explanation has been put forth.

Even if you seek a powerful experience, I strongly recommend that you take half of the intended dose, and wait an hour and a half before upping the amount. With psilocybin, there comes a point of diminishing returns; more is not always better.

Set and setting: the time and place for sacred moments

Set and setting refers to your mind-set and the physical environment in which the experience unfolds. You need to plan well in advance in order to maximize the benefits of set and setting. Set aside an entire day and evening. Your surrounding environment will affect you to the point that should you trip in a negative setting, the ensuing experience is likely to be terrible. Interestingly, I know of several people who have had nightmarish experiences but later went on to have tremendously rewarding trips once they had learned the importance of set and setting. Of course, there are many people who decide, after surviving a disastrous session, that they never want to trip again. Rather than blame themselves for improperly preparing for the experience, many will blame the mushrooms.

Obviously, those with heart conditions, kidney or liver disorders, or any other serious health problems would be ill-advised to experiment with psilocybin mushrooms. Likewise, mentally disturbed, violent, ill, or otherwise impaired individuals are not good candidates for tripping. Excessive doses exacerbate symptoms that are ordinarily not a problem. Some individuals feel a loss of strength and/or muscular control, drooping eyelids, an irregular heart beat, fever, deep sleep, and, rarely, convulsions. Some are frightened by what they see, and their body responds accordingly: accelerated heart beats, higher-than-normal blood pressure, and perspiration.

Before setting out on a mushroom session, users should come to an understanding with themselves: no matter what happens, you *will* come down. Do not panic. The experience will come, plateau, and gently descend. Those who panic guarantee that their experience will only become worse. The best assistance someone can provide to a person experiencing a bad trip is love, reassurance, and especially touch.

Physicians with whom I have tripped are watchful but do not seem alarmed, as these symptoms are only passing and recovery is typically complete. Recently, a psychiatrist who has enjoyed these mushrooms for years told me that in twenty years of medical practice, none of his patients has ever complained of a bad mushroom session. His list of individuals with bad reactions to legal and illegal drugs was long. In his professional point of view, psilocybin mushrooms are harmless in comparison. In fact, he feels that mushrooms can actually be beneficial for some patients, because of the self-examination process that often accompanies the experiences.

Trip with someone you love. This really helps. Tripping with strangers can be a major faux pas, and although you may build a friendship during the session, it is tough work compared to sharing intimate stories with a loved one. If this is your first experience, choose a partner who has successfully tripped before. Many psilonauts prefer taking their dog(s) with them. Canine friends can be an invaluable asset during an intense session. Dogs seem to know when you are tripping. Their heightened awareness and instinctive protectiveness make them great allies for sessions either at home or in the outdoors, especially if privacy is important to you and you want to be alerted to unexpected visitors.

Tell a confidant who is not part of the trip that you will be doing mushrooms. Tell them where you are going, with whom, and when you expect to be back. Lay out a plan, should you need to contact them for help of any kind. This confidant should also be an experienced veteran who understands that reassurance and calmness is the best prescription for anyone ill at ease with mushrooms.

The right setting is a highly subjective preference. The setting should be carefully reconnoitered beforehand, familiar, appropriate, and safe. I like to find a power spot—a place of natural beauty where base camp can be set up. I like being outside, high in the mountains, in the far reaches of canyons seldom traveled. I like views with great dimensionality. I love boulders, flowing waters, and open sky. And I love the centering effect of a campfire at base camp.

I usually trip with one, two, or three friends. We like places that project us into the panorama: vistas of mountains, valleys, waterfalls—all under star-speckled skies spanning across the horizon. From such vantage points, under the full force of mushrooms, the heavens open up with a display of beauty hard to describe. Visual acuity is enhanced to the point where the sky becomes three dimensional. The distances between stars and galaxies appear obvious. Electromagnetic fields ebb and flow, much like a hyperactive aurora borealis. On a recent trip, my eyes became so sensitive that, on a moonless night, I saw my hands cast shadows on the ground from the starlight above. Overlaying this display of splendor are colorful, dancing, geometrical fractals of infinite complexity. The universe moves in harmony. My spirit moves with it. I feel as though I have become a thread in the fabric of nature and have returned home. Experiences such as these leave impressions that are held dear for a lifetime. It is no wonder that cultures from Paleolithic times up to those of the present have all been held spellbound by psilocybin mushrooms.

I have had great experiences in remote mountain and beachfront cabins. The experiences began indoors, and as the session blossomed I have found that by forcing myself to get up and walk outdoors, the experience improved on many levels. Some people find themselves incapable of moving, practically paralyzed by the enormity of the experience. My suggestion: Get up! Move around! Hike on the beach or on a safe trail. If my footing is sure and the ground safe, I enjoy a brisk run. Then, upon returning to base camp, chanting and massaging my friends further embellishes the experience. Chanting becomes intimately tied to the visual realm, with cascading curtains and waves of geometric patterns unfolding and flowing as the harmonies intertwine. Once resonance patterns are in place, the mushrooms show their true power. Under these conditions, I have experienced unforgettable hallucinations: palaces of infinite grandeur, bejeweled and heavenly in nature, within a soaring emotional field of ecstasy and heightened intellect.

Like the shamans of southern Mexico, I prefer nocturnal tripping. I'll set up base camp in the late afternoon or early evening. An hour before sunset, the mushrooms are selected and separated into pairs. My wife and I have found that an offering of mushrooms to Gaia, on a makeshift altar, sets the stage for an experience filled with earth magic. Before ingestion, I like caressing their natural forms, speaking to them of their beauty, wisdom, and ancient power. They are the keys to the dimensions surrounding us that ordinarily cannot be seen. If they permit, you will be granted access to unimaginable dimensions of beauty, grace, and peacefulness. They bring me closer to God, Jesus, Buddha, Gaian consciousness, my origins, and to a deeper understanding of my purpose in the universe. The experience, by all measures, is profoundly spiritual. I strongly believe that the environmental movement that took off in the sixties has been and continues to be fueled by revelations from the psilocybin experience.

Four Mesoamerican mushroom stones, each nearly two thousand years old, in nocturnal contemplation.

CHAPTER 6

Field Collection Techniques

ORDINARY PEOPLE MORPH INTO strange forms of life when they take on the trappings of a sacred-mushroom hunter. The bizarre body language, erratic and stealthy, and the Latin words being spewed at every fungal encounter could cause the casual observer alarm. Even within the ranks of mushroom hunters, a range of distinct hunter dances can be observed. But edible mushroom hunters must totally relearn the art of hunting to find the psilocybin varieties. Since the psilocybin mushrooms are small in comparison to the common edible varieties, the technique for finding them is more subtle.

The species you are seeking will determine your technique for hunting them. The techniques of different mycophiles have always attracted my curiosity. Some less-athletic mycophiles I know insist that one of the better ways to hunt large, edible varieties (such as species of *Agaricus, Boletus,* and *Lepiota*) is to drive down country roads at no less than forty miles per hour, constantly glancing out the windows. When anything remotely resembling a mushroom is spotted, the car screeches to a halt and a scout is sent to investigate. This technique is hardly practical for finding most psilocybin mushrooms except the big Jim (*Gymnopilus spectabilis*) or golden tops (*Psilocybe cubensis*). For the better disguised, smaller candidates, different approaches have evolved. Many hunters prefer the random, fast-walk style of leaning forward at a precarious tilt. This insures forward momentum and a fast gait. Others are more systematic in their approach, slowly tracking along a mentally projected, predefined grid. Either method is measured by the success of the hunt at the end of the day.

Liberty cap (*Psilocybe semilanceata*) and blue veil (*Psilocybe stuntzii*) hunters have a very peculiar, off-balanced, forward-leaning posture, as they intently search the grass before them. Seasoned hunters often employ memory-mapping techniques—taking an image of the desired mushroom, mapping the image mentally, and then seeking a match in the background environment. When discovered, each mushroom virtually leaps from invisibility into the foreground. This method is the one I prefer. Once the first mushroom is spotted, I freeze and peripherally scan 360 degrees around the find. Although painstakingly slow, this method is highly rewarding, as I have picked several hundred specimens within just a few hours in this way. One note of caution: you can become so transfixed in your search that disorientation can occur. More than once, after looking up to see where I had wandered, I've felt waves of panic upon realizing I'd become totally disoriented.

Being transfixed can also cause retina burn. Retina burn is a neural phenomenon in which images of mushrooms, often species specific, appear when the eyes are closed. Sometimes persisting for hours, the images are often perfect in every detail and can even be studied. When a parade of mushrooms dances across your visual cortex, you are well on your way to becoming a true mycophile.

Some mushrooms can be targeted by zeroing in on a unique habitat. In Europe and in the Pacific Northwest, liberty caps, *Psilocybe semilanceata*, are concentrated in and around clumps of tall sedge grass in the low-lying wetlands of pastures.

A Mexican species, *Psilocybe muliercula* (= *Psilocybe wassonii*), can also be targeted in this way—it grows after high-mountain landslides. Interface environments, such as along borders where trees separate pastures, are often abundant with grassland and woodland species. Trees provide shade, humidity, and protection from the wind. They also tend to be places where cattle stand to stay cool.

Good field technique means collecting the mushrooms without damaging them. The loss of their identifying features, due to overhandling, complicates accurate identification. When picking a mushroom, be sure to get the entire fruiting body, especially the base of the stem where the bluing reaction is often first seen. If the base of the stem is not carefully preserved, sight identification of the mushroom is greatly jeopardized. I usually take my finger or a knife and carefully dig beneath the entire mushroom. Then, by gently lifting upwards, the mushroom can be picked fully intact.

Clean the debris off of each mushroom before it's placed into your basket. This will save you the inconvenience of dealing with the dirt that often covers the collection by the time you return home. Wrapping individual species in wax paper bags preserves the specimens until you have the opportunity to work on them. It also allows the mushrooms to breathe, whereas plastic tends to trap moisture and causes the mushrooms to quickly decompose. Finally, wax paper can help protect the mushrooms from being crushed by the weight of other specimens.

The traditional wicker basket is preferred by most collectors. Others favor a collapsible, epoxy covered basket, which is ideal if you are a mountain biker or hiker, as you can strap it on your back until it is needed.

Each collection should be labeled, and any notable features that might aid in identification recorded (such as bruising reactions, aspect, habitat, or nearby vegetation). When you find an unusual collection, thoroughly describe, photograph, and spore print it. (See pages 54–56.) The importance of spore printing all questionable and unfamiliar species cannot be overemphasized. Not only is the color of the spore print a primary key to identification, but for those skilled in mushroom cultivation, a clean spore print can be used for starting a culture. Also, dry, label, and save your specimens. Amateurs often find previously undescribed species. Most of the time they are unaware that their discovery is unique, and without an accurate description, proper classification is made very difficult.

Hunting for grassland Psilocybes in western Oregon, USA.

Twenty essential items for the sacred mushroom expedition

Any mushroom hunt could easily turn into a camping trip, so it's best to be prepared. The following items should be considered before you begin your hunt.

1. basket
2. wax paper
3. knife
4. maps
5. compass
6. first-aid kit
7. loud whistle
8. flashlight with an extra set of batteries
9. books—this identification guide as well as general guides to mushrooms in your area
10. water and food, especially fruits, nuts, and candy
11. matches, lighters, fire starter
12. good boots, extra socks, pants, gloves, hat, rainwear
13. environmental shelter, thermal blanket
14. money
15. extra set of car keys
16. cellular phone or handheld shortwave radio
17. GPS (ground positioning satellite) locator for those who are directionally challenged
18. insect repellent
19. camera and electronic flash
20. a positive attitude

Some items are more appropriate than others, and some may be better left in the car. This list will prove its usefulness over time.

Preservation and preparation of the mushroom sacrament

Mushrooms can be ingested using various methods. Once fresh mushrooms have been picked, they should be used within the day or dried within forty-eight hours. I do not recommend eating *Psilocybe* mushrooms raw, especially the ones growing in manured lands. They are often rife with maggots, small insects, fecal bacteria, and who knows what else. However, eating a *living* mushroom is an unparalleled method of making contact with *Psilocybe*.

I frequently ingest mushrooms in their dried form. This makes for a lot of chewing, and consumption of large amounts of water. Some people powder and encapsulate their mushrooms. This is unappealing to me, but I can understand why others like it—the dosage can be more accurately determined.

A better method is to make a mushroom tea by boiling water and adding prescribed amounts of mushrooms into the water while it is still hot. If you want to ingest them in a tea the same day they are picked, I recommend the following formula, known as the Magician's Brew:

The Magician's Brew: a mushroom elixir

We have developed mushroom formulas combining herbal teas, which impart to the brew a pleasant and soothing flavor. You will need two vessels. In one, add spiced, herbal tea—the best we have found uses a blend of orange peels, whole cloves, nutmeg, and cinnamon. Then add enough water to ensure the equivalent of one and a half cups per dose. Bring the water to a boil, then reduce the heat to simmer. Add the mushrooms in chopped or broken form into the bottom of the other empty vessel. Wrap the pot in a towel for insulation, then pour the hot tea through a strainer into the other pot. Stir occasionally. Allow to steep for an hour, and serve one cup per person per dose. After 10–20 minutes, cups of tea are distributed. At 1 gram of dried mushrooms or 10 grams of fresh mushrooms per cup of tea, the dosage can be accurately disseminated. Any residual mushroom debris is also evenly distributed and consumed.

The tea is only good for one day, after which ethyl alcohol (ethanol) must be added until the concentration reaches at least 70%. The high alcohol content is necessary to prevent fermentation. The brew should be refrigerated until use, at which time it should be warmed to evaporate off substantial amounts of alcohol. (Drinking a full cup of 70% alcohol is not advisable!) Although this preservation technique works well, the concoction will not last indefinitely. Bear in mind that a brew made from mushroom species higher in psilocybin than psilocin will have longer-lasting potency. When drinking the brew, be sure to swallow the leftover mushrooms.

Another method I have seen is to soak crushed mushrooms in 75+% ethanol. After two to three days, the roughage can be filtered, leaving a dark-blue elixir that can be decanted accordingly. For every fresh 5 grams of mushrooms, 25–50 milliliters of alcohol is recommended. Psilocybin and psilocin dissolve into this solvent, and the alcohol also acts as a preservative. I really don't have much experience with this technique, but have talked to people who say it is their preferred method. Some call this "blue juice."

For drying, food dehydrators, fans, and baseboard heaters all work well. For preserving the potency of psilocybin species, freeze-drying is the best method. However, freeze-dried mushrooms tend to crumble easily, and the process is not readily available to the general public. The next-best method is to thoroughly dry the specimens and then freeze them, sealed in airtight plastic bags. Drying the mushrooms in a food dehydrator works well only if the mushrooms are small. For larger mushrooms, the process is slowed, taking a day or two for complete dehydration. If the mushrooms cannot be thoroughly dried within two days, rotting is likely.

Although slicing large edible mushrooms is recommended for quickening the drying process, larger psilocybin mushrooms are better left intact. Cell-wall damage from cutting will result in a lowering of their potency. Since the larger psilocybin-containing species must be dried quickly, increasing the flow of dry air, as opposed to increasing temperature, is best. Temperatures should not exceed 120° F (50° C) For the more petite species, like the liberty cap (*Psilocybe semilanceata*), placing the mushrooms on a window screen several inches above a baseboard heater is probably one of the better methods—it dries the mushrooms in a matter of hours. Any system utilizing the flow of warm air through the specimens works well. Do not use solar dryers, or expose the mushrooms to direct sunlight, as ultraviolet radiation and extreme heat will reduce activity. Be careful about drying mushrooms above woodstoves, as the temperature fluctuates and is difficult to control.

Label each collection: species, date collected, and where it was found. Note any other pertinent data. After drying, seal them in an airtight plastic bag and freeze them. This will preserve their potency for the longest possible time. All species gradually lose their potency over time. Many species will lose most of their original potency after a few years of storage. *P. semilanceata* seems to degrade slowly, making it an excellent species to store over long periods of time. This is largely due to its relatively high psilocybin and low psilocin content. (See the chart at the top of page 40.) Psilocin is unstable compared to psilocybin.

Recent studies have also shown that *Psilocybe azurescens*, a new species from the Pacific Northwest, also degrades slowly, having approximately the same psilocybin and psilocin content after six months of storage as when it was fresh. However, most psilocybin mushrooms stored longer than a year usually show a significant loss of potency, especially in psilocin. The potency of species over the long term may also be a species-driven phenomenon. Some simply store better than others, for reasons not clearly understood. Another factor affecting the potency of Psilocybes is the condition of the mushrooms at harvesting. Older specimens, infested with parasites, will be less potent than younger specimens that were harvested in pristine condition. Specimens that have dried in the sun, are water soaked, have been frozen, or have become old vary in potency—complicating determination of dosage. However, many of us have found that, by mass, the juvenile mushrooms are usually much more potent than the adults.

The question of fasting before tripping frequently comes up. Many psilocybin users fast for a day in advance of the experience. I admire the discipline of those who do so, but I prefer to have breakfast and then nothing else the rest of the day until the session unfolds. During the trip, I often experience hunger but am not really hungry—a strange feeling that is readily satisfied by drinking a lot of water or juice coupled with some fruit or home-baked bread. Since the mushroom experience propels you into a fluid state of existence, drinking cold liquids feels especially satisfying.

CHAPTER 7

How to Identify Psilocybin Mushrooms

IDENTIFYING PSILOCYBIN MUSHROOMS to species is difficult unless you are a trained mycologist. Microscopic examination and comparison with known species can be a long and laborious process. However, determining whether or not a candidate is part of the psilocybin-containing group is not at all difficult. Again, I urge you to be cautious and not let your enthusiasm replace good judgment. Once you have successfully targeted a psilocybin mushroom, compare the mushroom in hand with the species descriptions listed in this book. When you are trying to identify a mushroom to species using macroscopic features, you are at best making an educated guess. Some species, like *Psilocybe semilanceata,* leave little room for confusion. Others may be difficult to pinpoint without closer study.

The Stametsian rule for targeting psilocybin mushrooms

How do I know if a mushroom is a psilocybin-producing species or not? If it matches these two conditions, it is a highly probable candidate:

> If a gilled mushroom has purplish brown to black spores, *and* the flesh bruises bluish, the mushroom in question is very likely a psilocybin-producing species.

I know of no exceptions to this rule, but that does not mean there are none! If you have a mushroom that bruises bluish but does not have purplish brown to black spores, there is a strong possibility it is not psilocybin. The bluing reaction is obvious in the more potent species, especially those high in psilocin. In general, the less psilocin there is in a species, the more subtle the bluing reaction. Be very careful, and *always,* before ingesting any mushroom, be sure of its identification. Always retain some specimens for later analysis, in case it becomes necessary.

The spore print

Most mushrooms continually drop spores from the time the cap expands, first exposing the gills. As the mushroom matures, increasing numbers of spores are let loose into the atmosphere until, at full maturity, spore production declines radically. In many cases, the color of the gills gives a hint as to true spore color. However, the color of the gills often does not reflect true spore color, due to the density of spore-producing cells and the color of the underlying flesh. Spores often collect on the stem, on any veil fragments (such as the annulus), and on adjoining shorter mushrooms. To the trained eye, spore color can be quickly surmised. Within most genera, the color of a fresh spore deposit is a fairly constant and reliable feature. The majority of distinctions between the gilled mushroom genera are at least partially based on spore color.

The collector must bear in mind that determining the color of spores is a subjective experience, thus there is margin for variation and interpretation. Perceiving the color of a spore print (for example, between a purple-brown spore print and a rusty-brown spore print) can be a distinction with deadly potential. Obviously, people who are color impaired should not be the ones making critical decisions about spore color.

Spore printing is an easy and fairly definitive means for separating groups of mushrooms from one another. *Psilocybe* and *Panaeolus*, genera that include the preponderance of psilocybin species, have spore deposits generally purplish brown to black in color. The previously mentioned genera that contain many toxic species have spore deposits with some shade of light brown, except for Amanitas, which are typically whitish. *Conocybe* and *Galerina* have rusty brown spores. *Entoloma* are pinkish brown. *Hebeloma* and *Inocybe* are yellowish brown to clay brown to dull brown.

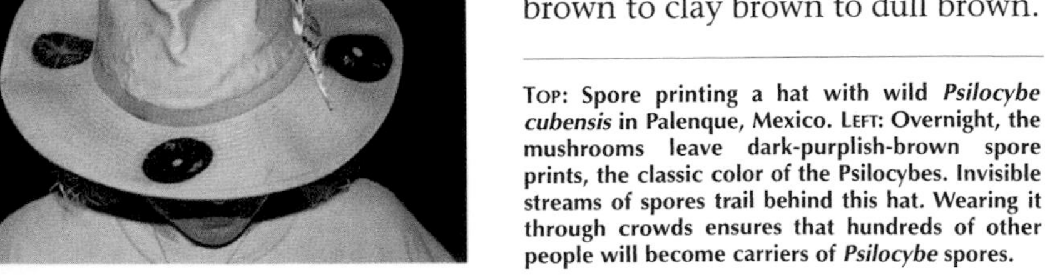

TOP: Spore printing a hat with wild *Psilocybe cubensis* in Palenque, Mexico. LEFT: Overnight, the mushrooms leave dark-purplish-brown spore prints, the classic color of the Psilocybes. Invisible streams of spores trail behind this hat. Wearing it through crowds ensures that hundreds of other people will become carriers of *Psilocybe* spores.

Although these are some of the more prominent hazardous brown-spored genera, keep in mind that an abundance of toxic species exists in other nonbrown-spored genera as well.

The technique for spore printing is very simple and is best done within the first few hours from the time the mushroom has been picked. As the mushroom dries, its production of spores declines, making printing more difficult. When collecting in the wild, first isolate the candidates and place them in a wax paper bag (not plastic—remember that the mushrooms have to breathe). When you return home, choose several fresh but fairly mature specimens. Specimens with convex caps are better than those that have flattened. With each mushroom, separate the cap from the stem, using a knife or scissors, and place the cap, gills down, on a piece of white paper or on a microscope slide. Placing a glass or cup over each cap to lessen the rate of dehydration and disturbance from air currents helps in obtaining a good spore print. Moistening the inside of the glass can also aid the spore-printing ability of partially dehydrated mushrooms.

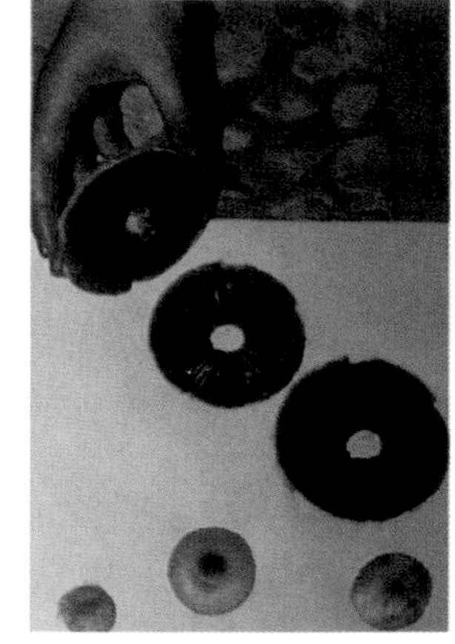

Spore printing wild *Psilocybe cubensis* on typing paper.

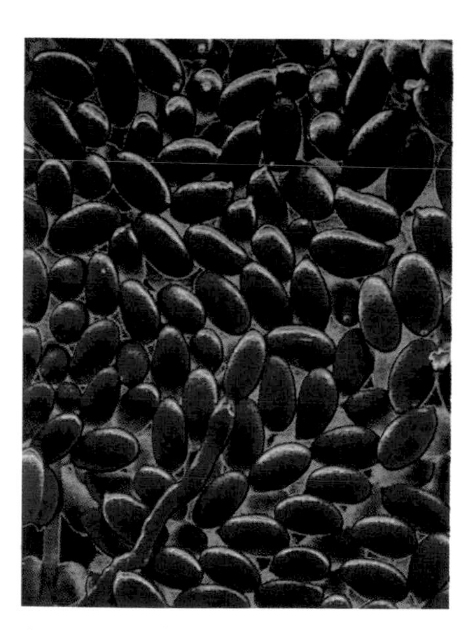

LEFT: Scanning electron micrograph, magnified approximately five thousand times, of the first of many *Psilocybe baeocystis* spores to germinate. RIGHT: Soon, the germinating spores generate a mycelial network, a biological "Internet" that collectively gives rise to mushrooms.

In a few hours, spores will deposit on the paper according to the radiating symmetry of the gills, indicating the spore color in mass. If you suspect that the spores will be white, take the print on dark paper. With the psilocybin species, white paper is recommended. Upon identifying the mushroom collection, label the print, seal it in a separate plastic bag, and save it with the original collection. If you wish to cultivate the species later, these printed spores can be used to establish cultures. Please consult my previous book *Growing Gourmet & Medicinal Mushrooms* for detailed instructions on the cultivation of mushrooms from spores.

The bluing reaction

A feature common to many of the psilocybin mushrooms is the bluing reaction. Many *Psilocybe* and *Panaeolus* species will turn bluish or bluish green when they're bruised. This happens either as a normal response to growing conditions, or while they are handled as they are picked. The bluing reaction is of great interest to chemists and pharmacologists. Even mushroom taxonomists consider it a valuable characteristic for delineating groups of species. Apparently, the blue pigmentation is a result of a phenomenon paralleling the degradation of unstable psilocin (dephosphorylated psilocybin) to presently unknown compounds by enzymes within the mushroom cells. What this means is that when a *Psilocybe* or *Panaeolus* bruises bluish, the color reaction is a co-indicator that psilocin is or was present. Naturally, since the bluing phenomenon appears to be a parallel decomposition sequence, the more the mushrooms are bruised the less potent they become. No one, to date, has been able to pinpoint the chemical structure of the bluing compound. Its elusive nature has been surprising to the chemists who have tried.

However, the bluing feature has limited importance from the taxonomist's point of view when it comes to identifying a mushroom. Many active *Psilocybe* and *Panaeolus* species will not blue no matter how much you abuse them, and there are several poisonous and suspect species outside these two genera that exhibit near bluing, although no psilocybin or psilocin is present. For instance, *Hygrophorus conicus* and allies turn brilliantly bluish black when disturbed, and they can be toxic, causing severe diarrhea. *Inocybe calamistrata* has a stem that is often blue black at the base, and although this species does not contain psilocybin, many poisonous, muscarine-producing Inocybes appear

***Psilocybe baeocystis* and *Psilocybe cyanofibrillosa* readily bruise bluish on the stem soon after being handled.**

virtually identical to the untrained collector. (Inocybes are difficult for most mycologists to identify with the naked eye.) I have also collected mushrooms belonging to the deadly genus *Galerina* whose stems have turned blackish from the base upwards. This blackish coloration could, by a stretch of the imagination, be called bluish black, especially at the stem base. Many mushrooms darken in this region. Don't be too eager to imagine a bluing reaction when, in fact, the mushroom is simply darkening from handling.

Although I have made every attempt to make this book as complete as possible, more psilocybin-active species are being discovered each year. Many of them have not been bioassayed. Several candidates are particularly mystifying: I view several polypores with suspicion, and a bluing *Lepiota* or *Hydnum* is curious to me. But of all the tantalizing genera, I am especially drawn to the small, delicate, but numerous Mycenas.

In the genus *Mycena,* which produces white spores, a half dozen species have bluish tones or turn bluish at the base. Mycenas are typically small, conic capped, woodland mushrooms. Despite repeated analyses (Klan 1985; Stijve and Bonnard 1986), no psilocybin or psilocin was detected in collections of *Mycena amicta, Mycena cyanorhiza, Mycena pelianthina,* or *Mycena pura.*

Nevertheless, I have heard several anecdotal accounts of activity from people ingesting Mycenas. John W. Allen and Gartz (1992) reports that a personal bioassay resulted in psilocybin intoxication on two occasions from two blue-footed Mycenas, a *"Mycena cyanorhizza"* and an unidentified *Mycena,* likely to be *M. amicta*. Because he is well versed in the effects of psilocybin, we should take his report seriously. However, we are faced with some contradictions.

I have been growing a luminescent *Mycena* from Malaysia, *M. chlorophos*, which produces white mycelium that bruises blue when cut. One analysis, however, have failed to detect any psilocybin, psilocin, or baeocystin. Is the bluing phenomenon seen in these *Mycena* chemically similar to that seen in the psilocybin-producing mushrooms? Or are the events totally unrelated? Many researchers agree that the bluing Mycenas should be broadly screened and compared to the psilocybin-active species. Until then, I do not recommend experimenting with these or any other mushrooms with which we have had little experience. Nevertheless, with the current state of knowledge, the bluing reaction is an especially good indicator of psilocin activity in *Psilocybe* and *Panaeolus*.

With these warnings taken into consideration, the bluing reaction can be a primary parameter for narrowing the field of mushrooms to the psilocybin-containing varieties with dark purplish brown to black spores. The bluing reaction is not very useful for determining the identity of individual species, nor is it useful for identifying all of the active species. However, the bluing reaction is an excellent, easy-to-observe feature that, in combination with other characteristics, particularly spore color, greatly narrows the field to a small pool of candidates.

Besides the spore print and the bluing reaction, some of the more representative characteristics of the genus *Psilocybe* are: they feature a separable gelatinous pellicle (see photo below), they have caps nut brown in color (which fade from the center to a straw color while drying—see the upper-left photo on the next page), their gill edges are usually fringed whitish (upper-right photo, next page), and the aspect is typically collyboid or mycenoid. The genus *Panaeolus* consists of primarily grassland- and/or dung-dwelling species, with gills that become spotted, with

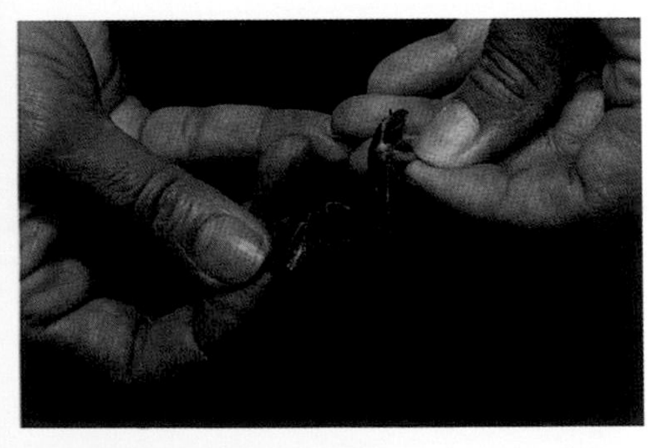

The separable gelatinous pellicle, a feature typical of *Psilocybe strictipes* and many Psilocybes, can be seen by carefully breaking the cap apart.

LEFT: In the genus *Psilocybe,* the cap color radically changes from brown to yellow or white as the cap dries, as seen here with *Psilocybe strictipes.* RIGHT: The whitish fringe along the gill edge can be seen clearly on many Psilocybes. This mushroom is a new *Psilocybe* species from northern Georgia. (See page 166.) BELOW: Many Psilocybes develop a fibrillose annular zone in the upper regions of the stem that is derived from the partial veil, and in this case (*Psilocybe azurescens*) is dusted with purple-brown spores.

maturity, with gray and black zones. The mushrooms are usually slender, with conic to hemispheric caps. The other genera hosting psilocybin mushrooms are discussed further on.

Before a mushroomer can effectively use this book and its keys, a certain understanding of terminology has to be reached. I do not submerge the reader into the sea of expressions mycologists often employ, but many terms cannot be substituted clearly or economically by common words without sacrificing precise shades of meaning. The language utilized here is not difficult to master. To make the transition easier, some of the more frequently encountered terms are listed in the glossary on page 203. I recommend spending some time familiarizing yourself with them, as well as the different kinds of cap shapes, gill attachments, and other general features illustrated in Diagrams A, B, and C on pages 200–202. After careful study, you should be fully prepared to use the information presented here to its greatest potential.

When identifying mushrooms, mycologists and amateurs commonly use taxonomic keys. The key on pages 62–64 functions through a process of eliminations. For the most part, it is a two-step (dichotomous) key that pairs distinguishing characteristics to arrive at species identification. By repeatedly selecting the key leads best fitting the appearance of the mystery mushroom, one is soon led to the identity of the mushroom genus. Since the key primarily uses macroscopic (visible) mushroom features, the reader must understand basic mushroom morphology. Most of the essential features are illustrated in Diagrams A, B, and C (pages 200–202), and others are described in the glossary. By taking spore prints and determining the spore color in mass to be purplish brown to black, and by checking this key, a person can key out the genus of their mushroom collection. If, when using the generic keys you are led to *Psilocybe,* or *Panaeolus,* consult the species descriptions for that genus. If your mushroom has a rusty brown, salmon-brown, or dull-brown spore print, identification becomes more problematic, as there are several deadly mushrooms with similar spore colors. However, if you have a mushroom with this spore color and it bruises bluish, please consult chapter 9, which includes the genera *Conocybe, Gymnopilus, Inocybe,* and *Pluteus.*

To best optimize identification using macroscopic features, a large collection of mushrooms is preferable. Fewer than three specimens will make identification extremely difficult. The mushroom collection should be representative, consisting of younger to older specimens. As mushrooms mature, their features evolve into different forms. This evolution of characteristics is of utmost importance. The most important observations, outside of habitat, include the progression of cap shapes from young to old, the manner by which the gills and stem meet, the transformation of the veil as it stretches from the cap margin to the stem, the changes in cap color when it's wet versus dry, the way the cap breaks apart upon bruising, and finally, to a lesser degree, odor and taste. In many cases, the use of a microscope will be necessary for absolute certainty. In some cases, the absence of distinguishing features—both macroscopic and microscopic—will make your identification little better than an educated guess. Again, I urge you to be cautious.

The following is an example of a typical narrowing-down technique.

> *Early one October morning, a friend and I set out hunting mushrooms in the rich, grassy fields around Stonehenge. Soon, we find an interesting and unfamiliar group of mushrooms growing in the tall clumps of grass. We pull out our list of key features and start the process of narrowing the field of candidates. We observe that they grow separately, but close together. We carefully pick the mushrooms with their stems intact, especially attentive to the fuzzy*

balls of mycelia at their bases. In general, the mushrooms have a small conic cap, a long stem, and are brown in color, with dark brown to purple gills. We collect about twenty specimens, including representatives of the youngest and the oldest. We wrap the find in twisted wax paper, label it ***Mushroom Collection X,*** *and include a small sample of the habitat.*

Hours later, we unwrap the collection and are startled at how much the mushrooms have changed. While drying out, the dark, rich, brown color is being replaced by yellow, drier tones. ***(The cap is hygrophanous.)*** *At the top of most caps are sharp nipples.* ***(The caps are acutely umbonate or papillate.)*** *The white mycelium at the base of the stem is struck through with blue tones. As the specimens dry further, more bluing is observed.* ***(The mycelium bruises blue.)*** *After separating the cap from the stem, we place the cap on a piece of white paper and cover it with a cup.* ***(The resulting spore print is clearly dark purple-brown to black.)*** *From a fresh, dark specimen, we slowly break the cap and notice a thin, translucent skin tearing off the surface as pieces are separated.* ***(The caps have a separable gelatinous pellicle.)*** *After a close look at the gills, we see that there is a fine, whitish band on the gill edges.* ***(There is cheilocystidia.)*** *Because of the presence of all these classic features, we conclude that the mushrooms belong to the genus* ***Psilocybe.***

We then set about limiting the sphere of possibilities to those psilocybin-containing mushrooms that grow in English pastures like the one near Stonehenge. We compare our collection with the species descriptions in this book, and again confirm that the spore deposit is purple brown, and that there is a bluing reaction. We feel sure that the mushrooms belong to ***Psilocybe semilanceata,*** *or the famed liberty cap, with* ***P. liniformans*** *and* ***P. strictipes*** *as close runners-up.) We disqualify* ***P. liniformans*** *because, when using a pin, we are unable to separate the whitish gill edge from the gill plates, a feature unique to that species. Most specimens in our collection have sharp nipples, features atypical of* ***P. strictipes,*** *hence that too is disqualified. This leaves* ***P. semilanceata*** *as the best possibility, given the spectrum of choices, and upon comparison of microscopic features, our identification is confirmed.*

The collection is given a date, and the location noted. Notes on the nearby flora, as well as soil types, are also taken. A general description, including measurements, further document the find, in the event that the mushrooms are someday deposited into a herbarium. Good photographs are taken.

This is just one approach to mushroom identification. Often, if you are unable to identify a mushroom, university botany departments (mycology laboratories) can help you. Some mycologists will happily identify a collection, but you should always ask them in advance before sending specimens. And please, as much as I would like to help you, do not send specimens to me for identification. I am, however, interested in seeing photographs of unusual species.

The following key can be used to determine the genus of mushroom species with brown to purplish brown or black spore deposits. By following it carefully, you will be led to the identification of the genus of the mushrooms at hand. The primary criterion is that the mushrooms you have found have dark brown to black spores. Once you have determined that your mushrooms fit this condition, simply follow the keys to see where they lead. Again, this does not apply to the rusty brown–, dull brown–, or salmon brown–spored mushrooms. Also, please note that this macroscopic key only *suggests* the mushroom genus, and works only for species exhibiting classic features. Atypical species will be more difficult to key out. Careful study of the mushroom, using both macroscopic and microscopic features, may be necessary for more accurate identification. Two exceptional species, whose identity will be difficult to key out using the above features, are *Panaeolus semiovatus* and *Psathyrella longistriata*. For these, please consult the individual descriptions in this book, as well as other field guides.

Generic key to agarics with dark-brown to black spores

1a Cap oysterlike, with an eccentrically attached, short stem: ***Melanotus***

1b Cap not oysterlike, with centrally attached stem: 2

2a Gills deliquescing (melting) into a black fluid, or becoming paper thin and disappearing at maturity: ***Coprinus***

2b Gills not as above: 3

3a Partial veil membranous (sometimes floccose), often leaving a membranous ring on the stem: 4

3b Partial veil not membranous, membranous ring not present: 10

4a Gills decurrent, thick, and waxy: 5

4b Gills not above: 6

5a Cap surface extremely viscid to glutinous when moist. Pellicle and partial veil thickly gelatinous, cap flesh whitish, not bluing in Melzer's iodine: ***Gomphidius***

5b Cap surface dry, sometimes viscid when moist. Partial veil not gelatinous. Cap flesh colored, bluing in Melzer's iodine: ***Chroogomphus***

6a Gills free. Spore deposit typically chocolate brown: ***Agaricus***

6b Gills attached (unless seceding). Spore deposit not as above: 7

7a Spore deposit typically black: ***Psathyrella***

7b Spore deposit not black, typically brown to purplish brown to very dark purplish brown: 8

8a Spore deposit typically dark purplish brown: ***Stropharia*** or ***Psilocybe***

8b Spore deposit typically dull to earth brown: 9

9a Spore deposit typically earthy brown. Cap usually smooth, and when moist splitting randomly when torn. (Cap cuticle cellular): ***Agrocybe***

9b Spore deposit yellowish brown. Cap often scaly or very viscid when wet and splitting radially when torn. (Cap cuticle filamentous): ***Pholiota***

10a Typically growing in dung, in well-manured grounds, or in rich grassy areas: 11

10b Typically growing in decayed wood, such as logs and stumps, or in wood chips and bark mulch: 14

11a Spore deposit blackish. Cap usually not viscid when moist and lacking a separable gelatinous pellicle: 12

11b Spore deposit purplish brown. Cap generally viscid when moist in most species and having a separable gelatinous pellicle: 13

12a Gills soon becoming spotted from uneven ripening of the spores when fully mature. (Spores not fading in concentrated sulfuric acid): ***Panaeolus***

12b Gills not becoming spotted. (Spores fading in concentrated sulfuric acid): ***Psathyrella***

13a Cap generally brownish when moist in fresh fruiting bodies and very hygrophanous (markedly fading in coloration upon drying). (Chrysocystidia always absent): ***Psilocybe***

13b Cap generally yellowish when moist in fresh fruiting bodies and usually not very hygrophanous (not markedly fading in coloration upon drying). (Chrysocystidia present): ***Stropharia*** or ***Psilocybe***

14a Cap generally colored in yellows, oranges, or reds when moist. Typically not very hygrophanous: ***Hypholoma***

14b Cap generally colored in deep browns when moist. Typically hygrophanous: 15

15a Cap usually not viscid when moist and lacking a separable gelatinous pellicle. Stem and cap never bruising bluish. Stem fragile, easily breaking. Cap brittle when squeezed, breaking randomly. (Cap cuticle cellular): ***Psathyrella***

15b Cap often viscid when moist from a gelatinous pellicle, separable in many species. Stem and cap sometimes bruising bluish. Stem tough, not easily breaking apart. Cap pliant, not truly brittle when squeezed, breaking radially. (Cap cuticle filamentous): ***Psilocybe***

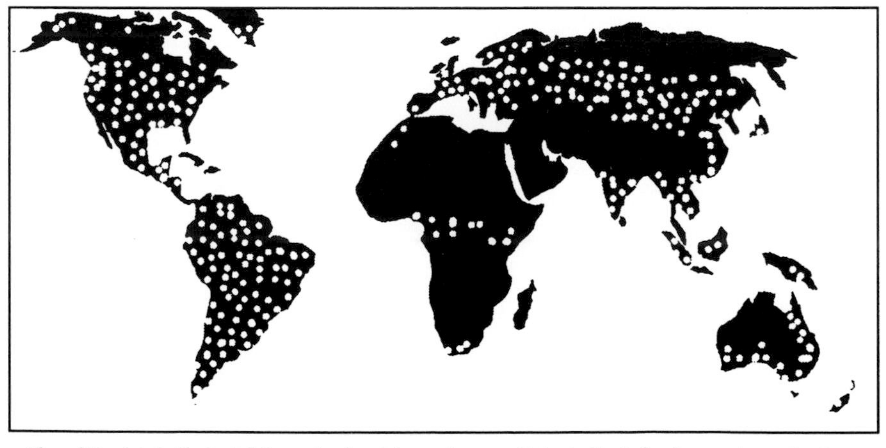

The white dots indicate sightings of psilocybin mushrooms. Refer to the Index for species endemic to countries and/or geographical areas.

THE MAJOR PSILOCYBIN GENERA

CHAPTER 8

The Major Psilocybin Genera

The genus *Panaeolus*

The genera *Panaeolus, Psathyrella,* and *Coprinus* belong to the family Coprinaceae. Most psilocybin species of this family are in *Panaeolus*. Species of *Panaeolus* are known by the mottled or spotted appearance of their gills just prior to being fully mature. This phenomenon is caused by uneven ripening of spore-producing cells (called basidia) on the gill surfaces. Many of the species in this genus grow in dung, while only a few Panaeoli are noncoprophilic, preferring grassy or, rarely, woodland habitats. The Panaeoli typically have hemispheric caps, relatively long stems, and produce black spore prints.

Psathyrella is quite similar in appearance to *Panaeolus,* but usually species of this genus grow in decayed wood substrata and in soils. The caps of Psathyrellas tend to be more striate than those of Panaeoli. Chemically, *Panaeolus* spores do not fade or discolor in concentrated sulfuric acid, while *Psathyrella* spores do. The Coprini are very distinct in the plicate (folded) nature of their caps and in that most species deliquesce at spore maturity (a process of autodigestion whereby the cap is reduced to a black liquid or becomes paper thin).

Following Dr. Rolf Singer's interpretation of this family, there are at least three other genera stemming from the traditional notion of the genus *Panaeolus*. Under the subfamily Panaeoloideae, he lists *Panaeolina* (nonannulate species with roughened spores, including *P. foenisecii* and *P. castaneifolius*.) *Copelandia* are tropical or semitropical Panaeoli that readily bruise bluish, and include the species *bispora, cambodginiensis, choloro-cystis, cyanescens,* and *tropicalis,*—all of which feature a characteristic form of pleurocystidia. (See the photographs on the following page.)

Anellaria include the Panaeoli, which are large, fleshy, often whitish with viscid caps. *P. semiovatus* (= *Anellaria semiovata*) is the foremost

LEFT: Photomicrograph of pleurocystidia typical of *Panaeolus cyanescens*. RIGHT: Scanning electron micrograph showing displaced pleurocystidia of *Panaeolus cyanescens* gill face.

species. The remaining species of *Panaeolus* not fitting into any of these categories are considered to be *Panaeolus* in the strictest sense, that is, *Panaeolus* sensu stricto.

The fractionation of such a closely knit, naturally allied group of species into separate genera seems artificial and unnecessary. Guzman and Perez-Patraca's (1972) treatment of the genus seems the most sensible to me: each of the above-named groups is a subgenera within an expanded concept of *Panaeolus*. Gerhardt (1987, 1996) further articulated the case for an expanded genus. Georges M. Ola'h, after spending a lifetime studying these fungi, currently believes the taxonomy of these species are better served under *Panaeolus* as one genus.

Therefore, for taxonomic clarity, I consider these genera or subgenera to be accurately represented under the epithet of one genus, *Panaeolus* sensu lato—*Panaeolus* in the broadest sense. Several species are consistent or latent producers of psilocybin and/or psilocin. Most all produce urea, serotonin, and tryptophan, and/or their precursors and derivatives (Stijve 1987). None have proven to be poisonous.

Panaeolus acuminatus (Schaeffer) Quelet sensu Ricken

= *Panaeolus rickenii* Hora

Cap: 1.5–2.5 (4) cm broad. Conic to conic-campanulate, becoming more campanulate to convex and in extreme age expanding to plane, often with a low umbo. Margin appendiculate with fine fibrils, if at all, shortly striate along the margin, then opaque, incurved when young and soon straightening. Cap chestnut or deep reddish brown when moist, hygrophanous, becoming more tawny in fading from the apex and remaining darker along the margin. Surface smooth, viscid to lubricous when wet, soon drying. **Gills:** Attachment adnate to adnexed to sinuate, close to crowded, broad, and even. Several tiers of intermediate gills present. Very dark purplish gray-black at maturity and mottled. **Stem:** 45–105 (150) mm long by 2.5–6.5 mm thick. Equal to slightly enlarged at the base or at the apex, tubular, and more or less brittle (cartilaginous). Very dark reddish brown or nearly concolorous with the cap. Surface pruinose and often with small water droplets adhering near to the apex of the stem. **Microscopic features:** Spores blackish in deposit, smooth, lemon shaped, 11–16 by 7.5–11 μ. Basidia 4-spored, rarely 2-spored. Cheilocystidia irregular in form. Pleurocystidia absent. **Habit, habitat, and distribution:** Grows scattered to gregariously in grassy areas and in well-manured grounds or on dung in the spring and fall throughout North America and much of temperate Europe. I often find this species along the field-forest interface. In the Pacific Northwest, this species is as common as *Panaeolus papilionaceus*. **Comments:** Not known to produce psilocybin and psilocin. Stijve (1987) found no psilocybin, no psilocin, .016% serotonin, .066% 5-OH-tryptophan, and .029% tryptophan. Gurevich (1993) did not detect any psilocybin or psilocin in a mushroom identified as *Panaeolus rickenii,* which most

Panaeolus acuminatus **is very common across the temperate regions of the world. It is not active.**

authors consider to be conspecific. Moser (1983) describes two forms, one as *Panaeolus acuminatus* ss. Ri., Lge, K. and R. (= *P. rickenii* Hora), which is campanulate when young, and another as *Panaeolus acuminatus* (Schff. ex Secr.) Quel., which has a dark band around the margin. The cap color and striations along the cap margin are distinctive features of this species. I often find beads of dew collecting on the upper regions of the stem, dusted with black spores. See also *Panaeolus foenisecii* and *Panaeolus castaneifolius.*

Panaeolus africanus Ola'h

Cap: 1.5–2 cm broad. Obtusely conic, hemispheric and rarely broadly convex in age. Surface smooth (but may crack to form scales when exposed to the sun), viscid when wet, especially in young specimens, and reticulated along the cap margin. Color grayish, creamy white, sometimes reddish brown towards the disc, and becoming grayish brown in age. Margin incurved when young, often irregular, and non-translucent. Flesh grayish white. **Gills:** Attachment adnate to adnexed, sometimes sinuate, rarely subdecurrent, widely spaced, irregular, grayish at first, soon grayish black, blackish with age, and mottled as spores mature. **Stem:** 30–50 mm by 4–6 mm thick, equal, firm, pruinose towards the apex. Whitish to white with pinkish tones, generally lighter than the cap, and lacking any veil remnants. **Microscopic features:** Spores nearly black in deposit, 11.5–14.5 by 7.9–10 μ, lemon shaped, and often variable. Basidia 2- and 4-spored. Cheilocystidia clavate,

Panaeolus africanus fruiting on hippopotamus dung. Weakly active.

17–24 by 7.4–12 μ. Pleurocystidia present, with extended sharp apices, 25–50 (60) by 10–17 (20) μ. **Habit, habitat, and distribution:** Reported from central Africa to the southern regions of the Sudan. Probably more widely distributed. Found on hippopotamus and elephant dung in the spring or during the rainy seasons. **Comments:** Contains psilocybin and psilocin, according to Ola'h (1969), although in irregular amounts. Years ago, I found this species on elephant dung from the Seattle zoo. I think the selling of "zoo-doo" to gardeners is one vector by which this species is spreading from the heartland of the African continent to other regions of the world. The robust *P. africanus* has a comparatively thick stem, a viscid cap, and is whitish in color. Visually, *P. africanus* is similar to *Panaeolus antillarum* (= *P. phalaenarum*). Microscopically, *P. africanus* has smaller spores than *P. antillarum*. Its penchant for hippopotamus and elephant dung allows this species to be easily identified. However, since many Panaeoli are highly adaptive, *P. africanus* may well spread to other ecological niches.

Panaeolus antillarum (Fries) Dennis

= *Panaeolus phalaenarum* (Fr.) Quelet
= *Panaeolus sepulcralis* Berk.

Cap: 4–10 cm broad. Hemispheric to broadly convex. Whitish at first and may become more yellowish towards the disc with age. Surface moist, smooth to rimose-scaly. Flesh relatively thick and whitish. **Gills:** Attachment adnexed to adnate close, broad, and slightly swollen in the

***Panaeolus antillarum*, a common but inactive mushroom.**

center. Color whitish to grayish at first, soon darkening to mottled blackish. **Stem:** 40–200 mm long by 5–15 mm thick. Equal to slightly enlarged and curved at the base, *solid,* and somewhat twisted. Whitish overall. Surface smooth to striate towards the apex. Context whitish. **Microscopic features:** Spores black in deposit, 18–22 by 11–12.5 μ, ellipsoid. Basidia 4-spored. Cheilocystidia cylindrical to clavate, 35–50 (80) by 6–12 μ. Pleurocystidia present, of the chrysocystidia type. **Habit, habitat, and distribution:** Gregarious to subcespitose in the spring or during the rainy season. Found from northern North America through Mexico into northern South America. Probably more widely distributed than presently reported. **Comments:** Edible but not high on my list of culinary delights. *Panaeolus phalaenarum* and *Panaeolus solidipes* (Peck) Sacc. are synonyms to *P. antillarum* (Fr.) Dennis. The form of this species that I have found in the northern latitudes tends to be much stouter and prefers horse dung, while the thinner forms found in warmer, subtropical zones thrive on cow manure. The larger, northern form generally resembles *Panaeolus semiovatus* except that this species boasts a well-formed membranous annulus, while *P. antillarum* is ex-annulate. Often, the Copelandian Panaeoli, like *Panaeolus cyanescens* and allies, look generally similar to the subtropical form of *P. antillarum* except they bruise bluish and are generally smaller in stature.

Panaeolus cambodginiensis Ola'h and Heim

= *Copelandia cambodginiensis* (Ola'h and Heim) Singer and Weeks

Cap: 1.2–2.5 cm broad, conic-convex at first, soon hemispheric and expanding to broadly convex, and eventually nearly plane. Surface smooth, moist to viscid when wet, soon drying, and often cracking with irregular, horizontal fissures. Young primordia are nearly chocolate brown in color, quickly fading in maturity to yellowish brown, often with olive greenish gray tones along the cap margin. Flesh pale but quickly bruising bluish upon injury. Margin incurved when young, often irregular, rarely with fine remnants of the veil, but not truly appendiculate. **Gills:** Pallid at first, soon grayish black to black with age and mottled from the uneven ripening of spores. Gill attachment ascending, uncinate, with several tiers of intermediate gills inserted. **Stem:** 55–95 mm long by 3.5–5 mm thick. Centrally attached, even, swelling towards the base, difficult to separate from the cap without gill fragments. Upper regions of the stem can be striate, below which a fine annular zone can sometimes be found.

Panaeolus cambodginiensis **fruiting on dung in Oahu, Hawaii. Potently active.**

Whitish to cream in color, brown near the base. Rapidly bruising bluish when injured. **Microscopic features:** Spores blackish brown in deposit, 10.5–12 by 6.5–9 μ, lemon shaped, smooth, with a relatively large, centrally located germ pore, dark brown under the microscope, nontransparent. Basidia 4-spored. Pleurocystidia present, fusoid-ventricose with a sharp elongated apex, measuring 48–60 by 13–19 μ. Cheilocystidia fusiform to clavate, 12–14 (20) by 2.5–5 (7.5) μ. **Habit, habitat, and distribution:** Scattered to gregarious, on the dung of water buffalo. Originally described from Cambodia. Thought to be widespread throughout the Asian subtropics. Merlin and Allen (1993) reported this species from Kahalu'u O'ahu, Hawaii. **Comments:** A strongly bluing species. Merlin and Allen (1993) reported the presence of psilocybin and psilocin, up to .55% and .6%, respectively. *P. cambodginiensis* has a golden-colored cap, especially towards the disc and, according to Ola'h (1969), is larger statured than its close relative, *Panaeolus tropicalis*, and generally smaller than *Panaeolus cyanescens*. However, I believe the composition of the dung greatly affects the size of the fruitbodies, and distinctions based on size have doubtful taxonomic significance. The late Steven H. Pollock was able to grow this species in a garden in Texas in the 1970s. From there, it has hopefully escaped into the Gulf Coast mycoflora. Another species, *Panaeolus bispora* Malencon and Bertault (= *Copelandia bispora* [Malencon and Bertault] Singer and Weeks), is virtually identical macroscopically to *P. cambodginiensis*, *P. cyanescens*, and *P. tropicalis*, except that it has exclusively 2-spored basidia. *Panaeolus bispora* has been reported from North Africa (Weeks et al. 1979) and from Hawaii (Merlin and Allen(1993).

Panaeolus castaneifolius (Murrill) Ola'h

Cap: 1–3 (4) cm broad. Distinctly campanulate at first, soon subhemispheric, then convex and becoming broadly convex in age. Margin incurved at first, soon straightening, not appendiculate, and slightly striated. Dark smoky gray when moist, hygrophanous, soon drying to a more straw yellow or pale ochraceous, and remaining more reddish brown at the apex and more smoky brownish along the margin. Surface sometimes finely wrinkled. **Gills:** Attachment adnate to adnexed close, and thin. Pallid at first, becoming dark purplish gray-black at spore maturity. **Stem:** 40–60 (75) mm long by 3–4 (6) mm thick. Equal to more narrow towards the base. Hollow or tubular, and brittle. Grayish to ochraceous or tan at the base. Surface slightly striated, pruinose. **Microscopic features:** Spores black in deposit, finely roughened, 12–15 by 7–9.5 μ. Basidia 4-spored. Cheilocystidia 20–28 (35) by 7–10 μ. Pleurocystidia few, or absent, not projecting beyond the plane of the basidia. **Habit, habitat, and distribution:** Growing scattered to gregariously in grassy areas across the North and South American continents. Possibly more widely distributed. **Comments:** Latently psilocybin, according to Ola'h (1969), and when so, weak. Some but not all collections of this species contain psilocybin. Distinguished from *Panaeolus foenisecii* by the color of the mature gills and spore deposit, which are very dark purplish gray-black.

LEFT: ***Panaeolus castaneifolius*** **grows in grassy areas. Inactive to slightly active. RIGHT:** ***Panaeolus cyanescens*****, highly active, is common on manure in the tropics and subtropics. This mushroom strongly stains bluish where bruised. Potently active.**

Panaeolus cyanescens Berkeley and Broome
= *Copelandia cyanescens* (Berk. and Br.) Sacc.
= *Copelandia papilionacea* (Bull. ex Fr.) Bres.

Cap: 1.5–3.5 (4) cm broad. Hemispheric to campanulate to convex at maturity. Margin initially translucent-striate when wet, incurved only in young fruiting bodies, soon opaque and decurved, expanding in age, becoming flattened and often split or irregular at maturity. Light brown at first, becoming pallid gray or nearly white overall with the center regions remaining tawny brown, soon fading. Cap cracking horizontally in age with irregular fractures and in drying. Flesh readily bruising bluish. **Gills:** Attachment adnexed, close, thin, with two or three tiers of intermediate gills. Mottled grayish black at maturity. **Stem:** (65) 85–115 mm long by 1.5–3 mm thick. Equal to bulbous at the base, tubular. Often grayish towards the apex, pale yellowish overall, then flesh colored to light brown towards the base; readily turning bluish when bruised. Surface covered with fine fibrillose flecks, which soon disappear. Partial veil absent. **Microscopic features:** Spores black in deposit, 12–14 by 7.5–11 μ, nontransparent, and without granulations. Basidia 4-spored, occasionally 2-spored. Pleurocystidia fusoid-ventricose, narrowing to an acute apex, 30–60 (80) by 12–17 (25) μ. Cheilocystidia present, 11–15 by 3–5 (6) μ. **Habit, habitat, and distribution:** Growing scattered to gregariously in dung in pastures and fields. In the United States, *P. cyanescens* can be found in Hawaii, Louisiana, and Florida. Widespread in most semitropical zones. Reported from Mexico, Brazil, Bolivia, the Philippines, eastern Australia, and occasionally in the Mediterranean region (near Menton, France). **Comments:** Moderately potent. A strong producer of psilocybin and/or psilocin, although highly variable in content. One study found .71% psilocybin, .04% psilocin, and .01% baeocystin. Virtually identical to *P. tropicalis* macroscopically, and differing only in the larger spore size and in the interior aspects of the spores. These subtle differences may be regional in nature, and I am not convinced these are distinct enough to warrant division. *P. cambodginiensis* is very similar, differing from the *P. cyanescens* in its overall size, habitat, and in the coloration of the apices of the pleurocystidia. I was once asked to identify an unknown mushroom near Tenino, Washington. It turned out to be *P. cyanescens*. The home owner had recently brought in manure to fertilize his yard from a local horse stable—which had just received horses from Florida. The mushrooms died out after two years.

Panaeolus fimicola **growing from cow dung. Not active to weakly active.**

Panaeolus fimicola Fries

= *Panaeolus ater* (Lange) Kuhner and Romagnesi

Cap: 1–2 (3) cm broad. Convex to campanulate, expanding to plane only during prolonged wet periods and in extreme age. Dingy gray to black, often with slight reddish hues, becoming paler in drying, leaving a darker grayish brown encircling zone along the margin. Margin striate when moist, not appendiculate, but sometimes irregular and uplifted at maturity. Surface smooth, soon dry. Flesh moderately thick. **Gills:** Attachment adnate to adnexed, broad, grayish, mottled, with grayish edges. Two to three tiers of intermediate gills present. **Stem:** 60–100 mm long by 1–2 mm thick. Equal, hollow, soft, and fragile. Dingy pale to whitish, especially towards the apex. Surface powdered above and striate below. **Microscopic features:** Spores black in deposit, 11–14 by 7–9.5 µ, lemon shaped with an apical germ pore. Basidia 4-spored. Pleurocystidia absent or only near to the gill edge and then similar in form to cheilocystidia. Cheilocystidia 25–35 (40) by 6–12 µ, mostly fusoid-ventricose. **Habit, habitat, and distribution:** Growing scattered in soil or dung from the late spring and in the fall. Sometimes found in well-fertilized lawns and/or grassy places in woods. Widespread, reported from the Americas, Africa, and Europe. **Comments:** Looking very much like a *Psilocybe*, especially in side view, this species is a latent producer of psilocybin, meaning that some collections, upon analysis, reveal a small amount of active indoles while other collections are devoid of them. Although widespread, this species is infrequently encountered compared to many other Panaeoli. Gerhardt (1996) showed synonomy between *Panaeolus fimicola* and *Panaeolus ater*. See also *Panaeolus castaneifolius, Panaeolus foenisecii*, and *Panaeolus papilionaceus*.

Panaeolus foenisecii (Fries) Kuhner

= *Panaeolina foenisecii* (Fries) Kuhner

Cap: 1–3 cm broad. Campanulate to convex with an incurved and occasionally translucent margin, especially in young specimens; expanding to broadly convex or nearly plane with age. Smoky brown to dull chestnut brown, hygrophanous, fading to sordid tan or light grayish brown, and often having a dark, ringlike band along the margin. Surface smooth, and often cracking in drying. Flesh thin, watery brown when moist, pallid when dry. **Gills:** Attachment adnate, and soon seceding, close, moderately broad, and slightly enlarged in the center. Color pallid in young fruiting bodies, darkening to a dull brownish or deep brown and becoming slightly mottled from the uneven ripening of spores. **Stem:** 40–80 mm long by 2–3.5 mm thick. Equal, brittle, pruinose, slightly striate and twisted towards the apex. Pallid to whitish overall, darkening from the base upwards with age and after being handled. Veil obscure or absent. **Microscopic features:** Spores dark brown, not black, in deposit, roughened, 12–17 by 7–9 μ. Cheilocystidia 22–30 (35) by 7–12 μ, variable, fusoid-ventricose, sometimes clavate or sublageniform with a long, flexuous neck. Pleurocystidia present in the form of chrysocystidia, but very difficult to find; generally not projecting beyond the basidia, 22–30 by 8–11 μ. **Habit, habitat, and distribution**: Scattered to gregarious in grassy areas but not

Panaeolus foenisecii **is especially common in yards in the spring through fall in temperate regions of the world. The spores of this mushroom are brown, not black as in other Panaeoli. Also, note the brown gills. Not active to weakly active.**

directly on dung. Found abundantly in the spring and to a lesser degree in the fall. Widely distributed throughout the temperate regions of North America, South America, and Europe. **Comments:** I do not believe this species is active. Beug and Bigwood (1982b) did not find any trace of psilocybin and/or psilocin in specimens from the Pacific Northwest. Ola'h (1969) reports some races are active. Notable features are its whitish striated stem, the occasional occurrence of a darker belt along the cap margin, and most importantly, the dark brown gills and dark brown (*but not black*) spore print. The roughened spores separate this species from most other Panaeoli. A form similar to *P. foenisecii,* which occasionally bruises bluish near the base of the stem, has been collected in grassy areas in coastal California. This mushroom may either be an active variety of *P. foenisecii* or is a new species. See also *Panaeolus castaneifolius.*

Panaeolus papilionaceus (Bull. ex Fries) Quelet

= *Agaricus callosus* Fr.
= *Panaeolus campanulatus* (Fries) Quelet
= *Panaeolus retirugis* (Fries) Quelet
= *Panaeolus sphinctrinus* (Fries) Quelet

Cap: (1) 2–5 (6) cm broad. Obtusely conic, becoming campanulate with age and having a margin decorated with whitish toothlike remnants of the partial veil. Occasionally with an obtuse umbo. Brownish to reddish brown, often with some grayish hues, and finally grayish cinnamon buff, while often remaining more tawny at the disc. Not particularly hygrophanous (fading in drying). Surface relatively smooth in young fruiting bodies and may be finely wrinkled to horizontally cracked in very mature specimens, exposing a lighter colored flesh, moist to smooth when wet but soon dry. Flesh thickest under the cap and thinning towards the margins; nearly concolorous with the cap. **Gills:** Attachment adnexed and soon seceding from the stem, close to subdistant, moderately broad, with 1–2 tiers of intermediate gills. Color grayish at first, becoming mottled dark grayish black from the uneven ripening of the spores. **Stem:** 60–140 mm long by 1.5–3.5 (5) mm thick. Equal, tubular, fibrous, and slightly striate towards the apex. Overall color brownish under a grayish pruinose surface. **Microscopic features:** Spores black in deposit, ovoid to lemon-shaped, 15–18 by 10–12 μ. Basidia 4-spored. Pleurocystidia absent. Cheilocystidia variable in form, from

***Panaeolus papilionaceus*—one of the most frequently encountered dung-inhabiting mushrooms in temperate climates. Not active.**

clavate to fusoid-ventricose with narrow apices, 18–26 by 5–8 μ. **Habit, habitat, and distribution:** Growing scattered to gregariously on dung in the fall or spring throughout North America and temperate regions of the world. Most common in the autumn months. **Comments:** Not known to be active. Undoubtedly, *P. papilionaceus* is one of the most commonly found of dung-inhabiting species of *Panaeolus*. Prominent for the toothlike veil remnants adorning the cap margin, *P. papilionaceus* has been in the center of a difficult-to-separate complex of mushrooms. In the past, there has been considerable confusion between *Panaeolus papilionaceus, Panaeolus campanulatus, Panaeolus retirugis*, and *Panaeolus sphinctrinus*, whose published descriptions share a strong overall resemblance. Gerhardt (1996) showed conspecificity of these previously separated taxa, with *P. papilionaceus* taking precedence. *P. papilionaceus* has a cap, which can become horizontally cracked in drying or in age. Prior to cracking, the cap can appear wrinkled, especially in young specimens. Whenever I find these species, I know that the environment is conducive for producing many of the psilocybin varieties, which I often discover growing in close proximity.

I have not felt any effect (other than malaise), even when I ingested nearly thirty specimens. Hence, I consider *P. papilionaceus* to be inactive. However, reports from early this century from Maine are enticing: two people ingested nearly a pound of mushrooms, and experienced a classic psilocybin experience (Verrill 1914). The mushrooms may have been incorrectly identified, and could have been *Panaeolus subbalteatus*. *P.*

papilionaceus gained notoriety when specimens rumored to be the sacred *teonanacatl* used in Aztec ceremonies were given in 1937 to Schultes, who placed them in the Farlow Herbarium at Harvard University, "doubtfully" identifying them as *Panaeolus campanulatus* var. *sphinctrinus* (Schultes 1939). The original collection was from Huautla de Jiminez, Oaxaca, Mexico. Later, although the specimens proved to be weakly active (Ola'h 1969, 270), these mushrooms were determined not to be the ones used by native peoples in their rituals. (Other species such as *Psilocybe caerulescens* and *Psilocybe mexicana* were favored.) This collection turned out not to be the much heralded "first" collection of *teonanacatl*. The mistake is understandable, considering the state of taxonomy at that time, and given that *P. papilionaceus* is a prominent dung-dwelling species. Another species, *Panaeolus alcidis*, which is smaller than *P. papilionaceus,* grows in Scandinavia and Canada on moose, reindeer, and roe deer dung. Moser (1984) described this species as similar to *Panaeolus sphinctrinus* var. *minor*, and differs in the absence of a partial veil.

Panaeolus semiovatus Fries (Lundell)

= *Panaeolus separatus* Gillet
= *Anellaria separata* Karst.

Cap: 3–6 (9) cm broad. Obtusely conic to parabolic at first, expanding to nearly convex. Cinnamon buff in young fruiting bodies becoming pinkish buff and fading in age to whitish. Surface viscid when moist, smooth to wrinkled. Flesh relatively thick, soft, and whitish. **Gills:** Attachment adnexed and soon seceding from the stem, close to subdistant. Color pallid to brownish and eventually mottled blackish from the uneven ripening of the spores. **Stem:** (80) 100–160 (180) mm long by (4) 6–10 (12) mm thick. Equal to slightly enlarged at the base; solid becoming tubular, and stuffed with a fibrous whitish pith. Whitish to pallid buff. Surface striate to smooth and powdered. Partial veil leaving a fragile membranous white annulus radially striate from the gills, and soon darkened by spores. **Microscopic features:** Spores blackish in deposit, ellipsoid, 18.5–21.0 by 10–11.5 μ. Basidia 4-spored. Pleurocystidia present, variable in form, of the chrysocystidia type, ampullaceous, 48–65 by 14–16 μ. Cheilocystidia 25–42 by 6–9 (12) μ, variable in form, fusoid-ventricose, pear shaped, clavate. **Habit, habitat, and distribution:** Solitary to scattered, primarily on horse dung in the spring and fall. Widely distributed throughout North America and temperate zones of

LEFT: The most majestic *Panaeolus, P. semiovatus,* grows on horse manure. Not active. RIGHT: *Panaeolus subbalteatus* is a distinctive mushroom also favoring horse manure. The band along the margin is characteristic of this species. Weakly to moderately active.

Europe. Also reported from Hawaii. Probably more widely distributed than reported in the literature. **Comments:** The most majestic of all the Panaeoli, there has been conflicting reports on the edibility of this species in earlier field guides. Most experts consider it nonactive and nonpoisonous. This species is distinguished by the membranous annulus, the viscid cap, and its relatively large size. I have only found this species on horse dung, especially around the compost piles created from cleaning out horse stables. See also *Panaeolus antillarum* (= *Panaeolus phalaenarum*).

Panaeolus subbalteatus Berkeley and Broome

= *Panaeolus venenosus* Murril

Cap: 4–5 cm broad at maturity. Convex to campanulate, then broadly convex, finally expanding to nearly plane with a broad umbo. Cinnamon brown to orange cinnamon brown, fading to tan in drying with a dark brown encircling zone around the margin. **Gills:** Attachment adnate to uncinate, close, slightly swollen in the center, and with three tiers of intermediate gills inserted. Color brownish and mottled, with the edges remaining whitish, blackish when fully mature. **Stem:** 50–60 mm long by 2–4 mm thick. Brittle, hollow, and fibrous. Reddish beneath minute whitish fibrils, darkening downwards. Oftentimes bruising bluish at the base. **Microscopic features:** Spores black in deposit, lemon shaped in side view, subellipsoid in face view. 11.5–14 by 7.5–9.5 μ. Basidia 2- and 4-spored. Pleurocystidia absent. Cheilocystidia variable in form, mostly pear shaped, 14–21 by 3–7 μ. **Habit, habitat, and distribution:** Grows cespitosely to gregariously in dung or in well-manured ground in the spring, summer, and early fall. Widely distributed. Reported from North America, South America, Europe, middle Siberia, Africa, and the Hawaiian archipelago (Ola'h 1969; Merlin and Allen 1993; Gurevich 1993). **Comments:** Weakly to moderately active. Stijve and Kuyper (1985) found maxima of .14% psilocybin, no psilocin, and .033% baeocystin. From Brazilian specimens, Stijve and Meijer (1993) found .08% psilocybin and no detectable psilocin. Gartz (1993) found maxima of 0.7% psilocybin, 0% psilocin, and 0.46% baeocystin. (However, there was nearly a tenfold variation in potency!) Gurevich (1993) found .14–.36% psilocybin from specimens collected in central Russia and middle Siberia. The cap and stem may bruise bluish over a long period of time, if at all, and most prominently in the mycelium attached to the base of the stem. I do not recommend ingesting this mushroom raw. Anecdotal reports from the Pacific Northwest associate the ingestion of raw mushrooms with stomach cramps, loss of muscular strength, and/or a feeling of malaise. However, Christian Ratsch found that this species often induces dreamy, aphrodisiacal experiences, and is increasingly popular in the resurgent pagan festivals of middle and northern Europe.

Other Panaeoli developing a dark band along the margin are *P. acuminatus, P. fimicola,* and *P. foenisecii.* However, the consistent dark band around the cap margin of *Panaeolus subbalteatus* is the most prominent feature distinguishing this species from all others. It is widely cultivated, unintentionally, on the discarded manure and straw from horse stables. Earlier this century, this "weed" mushroom caused several surprise intoxications, after it fruited in horse-manure compost at button farms (*Agaricus bisporus = Agaricus brunnescens*), and people assumed it too was edible. The tendency of the cap to expand to plane with age is also taxonomically significant. The synonym, *P. venenosus,* is a misnomer as no one has died from eating this mushroom.

Panaeolus tropicalis Ola'h

Cap: 1.5–2 (2.5) cm broad. Hemispheric to convex to campanulate. Margin incurved at first, may elevate slightly with age, and not strongly translucent-striate unless very wet. Pallid or grayish to yellowish brown towards the disc, hygrophanous, often developing a marginal zone, bluish in patches. Surface smooth to wrinkled, especially near the margin, and viscid when very wet. **Gills:** Attachment adnexed to more or less uncinate, subdistant, with several tiers of intermediate gills. Distinctly mottled, dull grayish with dark blackish spotted areas. **Stem:** 60–80 (120) mm long by 2–3 mm thick. Equal to swollen at the base, hollow. Grayish towards the apex, grayish brown in the middle area, and more blackish towards the base. Readily bruising bluish when touched. Surface longitudinally striate overall and pruinose towards the base. Partial veil absent. **Microscopic features:** Spores dark violet black to black in deposit, lemon shaped in side view, ellipsoid in face view, 10–12 by 7–9 μ. Basidia 2-spored. Cheilocystidia variable in form but mostly pear shaped, 18–30 by 8–10 μ. Pleurocystidia 45–55 (60) by 10–13 (14) μ. The spores of *P. tropicalis* are internally granulated, a feature that Ola'h (1969) used to delimit this species from *Panaeolus cyanescens*, its nearest relative. **Habit, habitat, and distribution:** Grows in cow dung and in the dung of wild animals in the tropics. Reported from Hawaii, central Africa, and Cambodia. The reports from southern California have not been verified by this author. **Comments:** Moderately active to potent. See also *Panaeolus cyanescens*, which has larger spores, and *Panaeolus cambodginiensis,* which has predominately 4-spored basidia rather than 2, and has pleurocystidia with much darker apices.

***Panaeolus tropicalis.* Potently active.**

The genus *Psilocybe*

Psilocybe mushrooms are strikingly similar to *Hypholoma (= Naematoloma*) and *Stropharia*. These three genera are difficult to separate from one another when relying solely upon macroscopic features. These naturally allied genera constitute the Strophariaceae family, which in its broadest interpretation includes the brown-spored genus *Pholiota* (Smith 1979; Singer 1975). *Psilocybe* mushrooms are saprophytes and can be found in a wide range of habitats: dungs, mosses, soils, grasslands, or decaying wood debris. When moist, most species have viscid, deep-brown caps that fade in drying to yellowish brown (i.e., hygrophanous). The more active species, particularly those high in psilocin, bruise bluish where injured. The gills are usually dark brown in color with whitish edges, and range from being subdecurrent to acutely ascending in their attachments. Guzman's monographs on *Psilocybe* (1983, 1995) are the most thorough to date, wherein he recognized 173 species. To date, the number of valid taxa approaches 180 species, with some authors proposing synonymies while others are further differentiating the genus into more taxa. At the time of this writing, about 95 species are thought to be psilocybin active, with more being discovered each year.

Taxonomic concepts are always in a state of flux and will continue to change as a new definition of the genus *Psilocybe* evolves. Many mycologists share the opinion that there should be one macrogenus, an expanded *Psilocybe* sensu lato, that would be all-inclusive of the species presently placed in *Stropharia* and *Hypholoma*. In the past, mycologists have been obligated to a system wherein *Psilocybe* is defined by its lack of the pigmented chrysocystidia cells found in *Hypholoma* and *Stropharia,* although other types of microscopic sterile cells on the gill are shared between all three genera. Often, a species that had been originally described as a *Stropharia* by one mycologist has been later classified as a *Psilocybe* by another. The renaming of *Stropharia cubensis* Earle to *Psilocybe cubensis* (Earle) Singer is a classic example. Most recently, Noordeloos (1995) took the bold, logical step of emending the definition of the genus so that most Stropharias will now be subjugated within an expanded concept of the genus *Psilocybe*. I wholeheartedly welcome this evolutionary step in taxonomy, and am following that emended definition in this book.

Psilocybe aeruginosa (Curtis: Fr.) Noordeloos

= *Stropharia aeruginosa* (Curtis: Fries) Quelet

Cap: 2–8 cm broad, convex to campanulate, soon expanding to broadly convex, often with a low, broad umbo. Dark bluish green, sometimes fading with maturity. Surface viscid when wet and covered with a bluish green, separable gelatinous pellicle. Margin even, adorned with whitish flecks, remnants from the partial veil. Flesh thin, whitish, thicker towards center. **Gills:** Adnately attached, broad, fawn to clay brown in color, sometimes tinged purplish, with white edges. **Stem:** 30–80 mm long by 3–12 mm thick, nearly concolorous with cap, equal, flared towards the apex, and swelled at the base. Surface covered with whitish patches. Partial veil membranous, leaving a fragile, membranous annulus on the superior regions of the stem, which is whitish above and bluish green below, sometimes disappearing in age. **Microscopic features:** Spores dark vinaceous purple brown to purple black in deposit, 7.5–9 by 4.5–5 μ, ellipsoid, thick-walled with a central germ pore. Cheilocystidia lageniform-capitate and bluntly capitate, 40–55 by 10–12.5 μ. Pleurocystidia mucronate, clavate, 40–60 by 10–15 μ. **Habit, habitat, and distribution:** Widespread throughout the British Isles, northern Europe, and in western North America, on wood debris, in gardens, parks, and occasionally along grassy areas at the edge of woodlands. In the Pacific Northwest this mushroom grows beneath conifers and in the South-

***Psilocybe aeruginosa*. Activity unknown.**

west under aspens. In southern California, it can be found under oak. **Comments:** Activity suspected but not known. A spectacularly beautiful mushroom, *P. aeruginosa* is listed in most books as a *Stropharia.* Noordeloos (1995) proposed a new combination, placing this mushroom, more appropriately, into the genus *Psilocybe,* following the suggestions of Alexander Smith (1979). This mushroom has historically been reported as poisonous, perhaps because of its psilocybin content. (Some books still report it is poisonous, without providing elaboration or references.) Analyses of specimens from Washington found no psilocybin or psilocin (Beug and Bigwood 1982b). I know of people who have eaten this species with no effects. In Europe, it is thought to be edible. Since the consumability of this species is questioned, caution is advised until the biochemistry of this species is studied further. See also *Psilocybe caerulea* (= *Stropharia caerulea).*

Psilocybe angustispora Smith

Cap: .2–.6 cm broad. Acutely conic to conic-campanulate. Margin decorated at first with whitish, fibrillose remnants of the veil. Dark reddish brown when moist, fading to pale pinkish tan in drying. Surface smooth and viscid from a thick gelatinous separable pellicle. Flesh very thin and pliant. **Gills:** Attachment subdecurrent, distant to subdistant, broad. Becoming dark purple-brown in age with the edges whitish fringed. **Stem:** 10–20 (40) mm long by .5–2 mm thick. Equal, flexuous. Pallid pinkish tan or nearly concolorous with the cap. Surface covered with pallid fibrils and the lower portions often with minute fibrillose scales. Partial veil thin, cortinate. Base usually adorned with tufts of mycelium. **Microscopic features:** Spores dark purplish brown in deposit, narrowly ellipsoid, 12–15 by 5–8 μ. Basidia 4-spored. Pleurocystidia absent. Cheilocystidia present. **Habit, habitat, and distribution:** Single to several on the dung of sheep, cows, horses, elk, marmots, and other animals during the spring and fall. Originally, reported from western Washington, Oregon, Idaho, and Colorado. Probably more widely distributed. **Comments:** An uncommon species in need of further study, this petite species is a classically shaped mycenoid *Psilocybe.* I suspect that it is active, but no one has yet submitted specimens for analysis. See also *Psilocybe semilanceata* and allies.

Psilocybe argentipes Yokoyama

Cap: 2.5–5 (6) cm broad. Conic to conic-campanulate at first, soon expanding to convex, broadly convex, and eventually plane in age, and variably with a sharp umbo. Chestnut brown when wet, hygrophanous, fading in drying from disc to a golden brown, honey yellow or mustard brown or clay toned, and bruising bluish. Margin incurved when young, irregular and often wavy, and adorned with remnants of the partial veil, especially when young. Surface smooth but not viscid when moist. **Gills:** Attachment adnate or adnexed, soon seceding. Grayish orange at first, eventually purple brown with whitish edges. **Stem:** 60–80 mm long by 2–4 mm thick. Equal, enlarging towards the base from which whitish rhizomorphs radiate. Silky white at first, then yellow toned, soon brownish to reddish brown, and adorned with white fibrillose veil remnants in the lower two thirds of the stem, yellowish brown, and whitish near the apex. Partial veil cortinate, white, and leaving a fragile annular zone on the stem, if at all, and soon disappearing. **Microscopic features:** Spores dark violet brown in deposit, subellipsoid, 6.5–7.5 by 9.5 by 3.3–4.4 μ. Basidia 4-spored. Pleurocystidia absent. Cheilocystidia 13–25 (32) by 5–8 μ. **Habit, habitat, and distribution:** Gregarious to

Psilocybe argentipes. Note purple-brown spores collecting on the caps. Moderately to highly active.

clustered on soil rich in woody debris, along trails, underneath or nearby *Cryptomeria japonica, Quercus glauca,* or *Pinus taeda.* Known only from Japan. **Comments:** Active, according to Koike et al. (1981). This species is named for the silvery patches of fibrils adorning the stem. *P. argentipes* is likely to be fairly potent—comparable to *Psilocybe cyanescens.* This *Psilocybe* may be distributed outside of Japan. See also *Psilocybe azurescens, Psilocybe caerulipes, Psilocybe cyanescens, Psilocybe cyanofibrillosa, Psilocybe muliercula,* and *Psilocybe suberuginosa.*

Psilocybe atrobrunnea (Lasch) Gillet

Cap: 2–4 (6) cm broad. Bluntly conic to convex or campanulate, usually umbonate, sometimes with a sharp nipple, expanding to broadly convex in age. Dark reddish brown to blackish reddish brown, then brown, strongly hygrophanous and fading to pale reddish brown in drying. Surface smooth, translucent-striate near the edge, viscid when moist from a thin but often separable gelatinous pellicle. Margin inrolled to incurved at first, and adorned with whitish veil remnants. **Gills:** Attachment adnate to adnexed, dull cinnamon brown to dark purplish brown at maturity, with whitish edges, sometimes irregular. **Stem:** 80–180 mm long by 3–5 (6) mm thick. Equal, tough, flexuous, swelling towards the base, which is adorned with whitish mycelium. Reddish to blackish underneath the appressed, fibrillose whitish remnants in the lower two thirds, and pruinose above. **Microscopic features:** Spores dark violaceous brown in deposit, ellipsoid, 9–12 (14) by 5–7 (9) µ. Basidia 4-

Psilocybe atrobrunnea. Not active to weakly active.

spored. Pleurocystidia absent, or close to gill edge, and then similar to cheilocystidia, 30–38 by 4–6 (8) μ. Cheilocystidia lageniform or fusoid-ventricose with an extended neck, 18–30 (36) by 4–7 μ, and 1.5–2.5 μ thick. **Habit, habitat, and distribution:** Gregarious to scattered in or near sphagnum bogs, in coniferous and deciduous woodlands, fruiting in September and October. Reported from the United States (Michigan and upper New York to Maine), as well as British Columbia and central to northern Europe (Great Britain, Czech Republic, Slovakia, Finland, France, Germany, Sweden, and Poland). Probably more widely distributed. **Comments:** Possibly active. Høiland (1978) reported psilocybin activity from specimens collected in Norway. No other studies are known to me. This species has an overall resemblance to *Psilocybe washingtonensis, Psilocybe physaloides, Psilocybe inquilina,* and particularly to several Hypholomas, namely *H. dispersum* (= *Naematoloma dispersum*) and *H. udum* (= *Naematoloma udum*). Its fondness for sphagnum bogs makes that habitat target-specific.

Psilocybe aucklandii Guzman, King and Bandala

COMMON NAMES: the Auckland *Psilocybe,* King's *Psilocybe*

Cap: 1.5–5.5 cm broad. Broadly conic, expanding to broadly umbonate, becoming nearly plane in maturity. Margin striate, upturned, irregular, and splitting in age, free of veil remnants. Dark brown to yellowish brown, hygrophanous, drying to pale yellowish brown to straw colored. Flesh white, bruising bluish in age or where injured. **Gills:** Attachment adnate, grayish yellowish brown, darker, with maturity. Edges whitish. **Stem:** 35–100 mm long by 1.5–5 mm thick, even, pruinose above, covered with whitish, silky fibrils below. Flesh brownish, bruising bluish. Partial veil cortinate, poorly developed, soon disappearing. **Microscopic features:** Spores dark purplish brown in deposit, ovoid-ellipsoid in side view, ovoid in face view, (6.5) 7–9.5 by 3.5–5.5 μ. Basidia 4-spored. Pleurocystidia 13–19 by 4.5–6 μ, scattered, similar to cheilocystidia but with a shorter neck, up to 4.5 μ. Cheilocystidia 15–32 by 4–8 μ, ventricose-rostrate, with a long flexuous neck up to 12 μ long, often forking. **Habit, habitat, and distribution:** Common. Grows scattered to gregariously on soil rich in woody debris and litter beneath *Leptospermum* and *Dacrydium,* and in pine (*Pinus radiata*) plantations. Reported only around Auckland, New Zealand. **Comments:** Estimated to be moderately potent. No analyses are

Psilocybe aucklandii.

known to me. With the heavy export of lumber and raw logs from the pine plantations of New Zealand, this species has a direct gateway for spreading to many other temperate regions of the world. This species resembles many of the lignicolous Psilocybes. See also *Psilocybe makarorae, Psilocybe subaeruginosa, Psilocybe cyanescens, Psilocybe cyanofibrillosa,* and *Psilocybe bohemica*.

Psilocybe australiana Guzman and Watling

Cap: 1.5–3 cm in diameter, convex to subcampanulate, slightly umbonate, and sometimes with a short, sharp nipple. Surface smooth, viscid when wet, often decorated with veil remnants along the cap margin, especially when young. Light orangish brown, darker towards the center, fading in drying to yellowish brown, and often bluing along the cap margin. **Gills:** Attachment adnate, olive yellow at first, eventually dark purplish brown with maturity. **Stem:** 45–110 mm long by 2–3 mm thick. Equal overall but swelling towards the base. White to dingy brown in color, bruising bluish when handled. Partial veil cortinate, white, leaving a fragile, annular zone often dusted with dark spores. **Microscopic features:** Spores dark purplish brown in deposit, subellipsoid to ovate, 12–14 (16) by 6–8 μ. Basidia 4-spored, rarely 2-spored. Pleurocystidia 22–33 by 8–9 μ, lageniform, hyaline, common near to the gill edge and infrequently dispersed elsewhere.

Habit, habitat, and distribution: First reported from southern Australia (New South Wales), near Sydney, and also known from Tasmania. Preferring grounds rich in woody debris from *Pinus radiata* and on twigs and branches beneath *Eucalyptus* and *Nothofagus* trees. Found in April along road sides, in tree plantations, along trails, and similar habitats. **Comments:** A mildly potent species, this mushroom is widely sought after by amateur mycologists. The extensive, commercial timber plantations using *Pinus radiata* have created an extensive ecological zone in which this species flourishes. More recently, this mushroom is becoming increasingly associated with gardens, thriving in the debris fields around homes.

A controversial article by Chang and Mills (1992) sought to show synonymy between *Psilocybe australiana*, *P. eucalypta*, and *Psilocybe subaeruginosa* based on isozyme analysis, intercompatibility of spores, and electrophoretic data (zymograms). Although intercompatibility between the listed collections was reported, the authors noted that they did not compare compatibility with the type collection of *P. australiana,* as it could not be accessed. Furthermore, their wild collections of *Psilocybe subaeruginosa* failed to have the species' most unique feature—brown pleurocystidia. These two uncertainties cast doubt, in my mind, about their case for conspecificity. The other similarities they reported are not surprising, given the tight taxonomic relationship of

Psilocybe australiana. **Moderately active.**

these three taxa. Guzman et al. (1993) took issue with Chang and Mills' results. Until further studies can be juried, I am following Guzman and Watling's (1978) delineation of these entities into three discrete taxa. Nevertheless, the opposing points of view underscore the tight taxonomic relationship shared by these close relatives. They may represent a cluster of varieties with recent ancestry in the process of active evolution, differentiating away from one another. Besides those mentioned above, see also *Psilocybe cyanescens*, *Psilocybe mairei*, *Psilocybe serbica*, and *Psilocybe azurescens*.

Psilocybe aztecorum Heim emend. Guzman

COMMON NAME: ninos (*apipltzin*, or "little children" in Nahuatl)

Cap: 1.5–3.5 cm broad, obtusely conic or campanulate at first, soon convex, expanding to broadly convex, plane or even uplifted at full maturity. Sometimes the center can be depressed into the stipe, an occasional deformity. Surface viscid when moist, smooth, translucent striate along the margin. Dark chestnut brown, strongly hygrophanous, fading to straw yellow to off white in drying, often bruising, especially along the cap margin, which is typically even. **Gills:** Attachment adnate to adnexed, light purplish gray to dark purplish brown, with whitish edges. **Stem**: (25) 55–75 (95) mm long by 3–4 mm thick. Equal but thicker at the base and towards the apex, straight to flexuous, pruinose above and silky-fibrillose below. White rhizomorphs radiate from the base of the stem. Whitish to grayish, easily bruising bluish. Partial veil cortinate, white, soon disappearing, and occasionally leaving an annular zone in the upper reaches of the stem. **Microscopic features:** Spores violet black in deposit, 10.5–14 by 7–9 μ, ellipsoid. Basidia 1-, 2-, 3-, or 4-spored. Pleurocystidia few, scattered, same as cheilocystidia. Cheilocystidia 20–45 by 5–8 μ, fusoid shaped with a long, thin neck 6–11 by 1.5–2.5 μ.

Habit, habitat, and distribution: Numerous to gregarious, on soil rich in wood debris, rarely on pinecones, in open woods of *Pinus hartwegii* rich with grasses. Fruiting from August through October in the high mountains of central Mexico at 3200–4000 meters above sea level. This species is probably more widely distributed than reported. **Comments:** Fresh specimens strongly bluing and potent. Heim and Hofmann (1958) detected only .02% psilocybin and no psilocin, not surprising since the analyses were conducted two years after the mushrooms were originally collected. *P. aztecorum* is one of two probable candidates for the *teonanacatl* that was reported by Sahagun in the sixteenth century. This woodland species shares many characteristics with the more temperate Psilocybes such as *P. baeocystis*, *P. quebecensis*, and similar species. In Guzman's (1983) revision of the species, he recognized two varieties, *P. aztecorum* var. *aztecorum* and *P. aztecorum* var. *bonetii*, the latter preferring a lower altitude (2000–3300 meters) and associated with pine, *Pinus montezumae*, and fir, *Abies religiososa*.

Above and facing page: ***Psilocybe aztecorum*. Moderately to highly active.**

Psilocybe azurescens Stamets and Gartz

COMMON NAMES: Astoriensis, flying saucer mushroom, indigo *Psilocybe,* blue runners, blue angels

Cap: 3–10 cm broad, conic to convex, expanding to broadly convex and eventually flattening with age with a pronounced, persistent broad umbo; surface smooth, viscid when moist, covered by a separable gelatinous pellicle; chestnut to ochraceous brown to caramel in color, often becoming pitted with dark blue or bluish black zones, hygrophanous, fading to dingy brown in drying, strongly bruising blue when damaged. Margin even, sometimes irregular and eroded at maturity, slightly incurved at first, soon decurved, flattening with maturity, translucent striate and often pale azure tinted. Flesh at center 3–6 mm thick, cottony and whitish to pallid brown at the stipe/pileus junction before bruising. Flesh rapidly bruising blue, then darkening to deep caerulean blue and eventually indigo-black. **Gills:** Attachment ascending, sinuate to adnate, brown, often stained indigo black where injured, close, with two tiers of lamellulae, mottled, edges whitish. Spore print dark purplish brown to purplish black in mass. **Stem:** 90–200 mm long by 3–6 mm thick, silky white, dingy brown from the base or in age, hollow at maturity. Composed of twisted, cartilaginous, silky white fibrous tissue. Base of stem thickening downwards, often curved, and characterized by coarse white aerial tufts of mycelium, often with azure tones combined with dense, thick, silky white rhizomorphs that tenaciously attach to

Psilocybe azurescens **thrives in a combination of wood chips and tall grass. Note that low umbo and evenness of cap margins are consistent features. Extremely potent.**

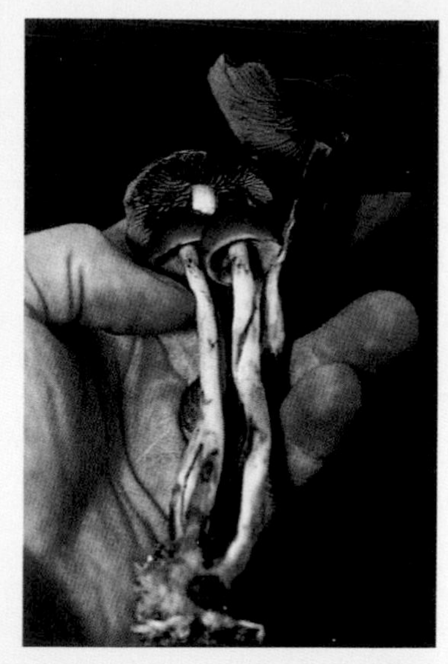

LEFT: ***Psilocybe azurescens.*** Note that remnants of the partial veil have left an annular zone on the stem, which has become dusted with purple-brown spores. RIGHT: ***Psilocybe azurescens,*** one of the most potent species in the world, exhibits an unusual bluing reaction due to its high indole content: the flesh can become indigo-black from bruising.

wood chips or dead grass, strongly bruising bluish upon disturbance. Veil white, cortinate, often leaving a fibrillose annular zone in the superior regions of the stem. Taste is extremely bitter. Odor none to slightly farinaceous. **Microscopic features:** Spores dark purplish black in deposit, 12–13.5 by 6.5–8 μ, ellipsoid. Basidia 4-spored. Pleurocystidia abundant, fusoid-ventricose, tapering to a narrow but short neck, bluntly papillate, 23–35 by 9–10 μ. Cheilocystidia nearly identical to pleurocystidia, measuring 23–28 by 6.5–8 μ. Tissue notably awash with bluish tones. **Habit, habitat, and distribution:** Cespitose to gregarious on deciduous wood chips and/or in sandy soils rich in lignicolous debris. This mushroom naturally grows, often prolifically, along the northern Oregon coast near Astoria, favoring the land adjacent to the shoreline. *P. azurescens* has a strong affection for dune grasses, especially *Ammophila maritima,* with which it is closely associated. Generating an extensive, dense, and tenacious mycelial mat, *P. azurescens* causes the whitening of wood. Fruitings begin in late September and continue well after the first frost, often into late December and early January. An adaptive species, outdoor beds have been established with ease in the United States (California, New Mexico, Wisconsin, Vermont, and Ohio) and Germany (Leipzig). **Comments:** Extremely potent, containing up to 1.78% psilocybin, .38% psilocin, and .35% baeocystin (Stamets and Gartz 1995). After six months of storage, analyses revealed that this species retained most of its original potency. This species is unique not only in its potency, but also in its relatively high baeocystin content, roughly on par with *Psilocybe semilanceata*. A cold weather–tolerant species, *P. azurescens* is one of the most potent species in the world and exhibits one of the strongest bluing reactions I have seen. The flesh actually becomes indigo black where damaged. (See upper-right photo, previous page.) The silky white stem, caramel-colored cap, relatively large stature, nonundulating cap margin, and broad but pronounced umbo are representative features. *Psilocybe cyanescens* is similar to *P. azurescens* but can be macroscopically distinguished from it by its much smaller stature, and characteristically sine-wave margin. A source on the Internet lists a mushroom from Astoria, Oregon, called *"Psilocybe cyanescens* Ossip," which I recently viewed. The mushrooms bore a resemblance to *P. azurescens*. See also *Psilocybe cyanofibrillosa, Psilocybe serbica, Psilocybe subaeruginosa,* and *Psilocybe venenata.*

Psilocybe baeocystis Singer and Smith

COMMON NAMES: baeos, knobby tops

Cap: 1.5–5.5 cm broad. Conic to obtusely conic to convex, expanding to plane only in extreme age. Margin incurved at first, and distinctly undulated when convex; translucent striate and often tinted greenish. Dark olive brown to buff brown (occasionally steel blue), becoming copper brown in the center when drying, hygrophanous, fading to pallid white and easily bruising bluish. Surface viscid when moist from a gelatinous pellicle, usually separable. **Gills:** Attachment adnate to sinuate, close. Color grayish to dark cinnamon brown with the edges remaining pallid. **Stem:** 50–70 mm long by 2–3 mm thick. Equal to subequal. Pallid to brownish surface sometimes covered with fine whitish

Psilocybe baeocystis **prefers bark mulch or wood-enriched soils under conifers. Moderately to highly active.**

Psilocybe baeocystis **fruiting through a moss island with wood buried underneath.**

fibrils, while often more yellowish towards the apex. Brittle, stuffed with loose fibers. Distinct rhizomorphs present about stem base. Partial veil thinly cortinate, rapidly becoming inconspicuous. **Microscopic features:** Spores purplish brown, mango shaped, 10–12 by 6–7 μ. Basidia 4-spored. Pleurocystidia absent. Cheilocystidia 20–30 (40) by 4.5–6 (9) μ, fusoid with a narrow neck. **Habit, habitat, and distribution:** Solitary to gregarious to subcespitose on decaying conifer mulch, in wood chips, or in lawns with high lignin content. Occasionally grows from fallen seed cones of Douglas fir. Found in the fall to early winter and rarely in the spring. (I once found it as early as June 20.) First reported from Oregon, common throughout the Pacific Northwest. **Comments:** Moderately active, losing significant potency in drying or from damage. *P. baeocystis* ranges .15–.85% psilocybin, up to .59% psilocin, and up to .10% baeocystin (Repke et al., 1977; Beug and Bigwood 1981; Beug and Bigwood 1982b).

McCawley et al. (1962) first analyzed mushrooms purported to be *P. baeocystis,* finding .63% psilocybin and <.1% psilocin. Their research was spurred by the unintended intoxication in 1960 and 1961 of eight individuals. One child died after three days of hospitalization, subsequent to running a 106° F fever. An autopsy showed cerebral edema, a symptom I note that is consistent with *Galerina* and *Amanita* mushroom poisonings. Examination of the photographs of the alleged toxic species clearly show mushrooms resembling *Psilocybe cyanescens*, not *P. baeocystis*. In conversations I had with the identifying mycologist, Alexander

Smith, and upon being confronted with the apparent discrepancy of the photographs—which he had never seen—he retracted his identification on the spot. Furthermore, he said, mushrooms other than Psilocybes were present in the yard of the fatal child, but were not kept. The *Psilocybe* was. In this case, Dr. Smith told me he suspected that several species were involved. The record needs to be emended, as *P. baeocystis* is erroneously attributed to this report. The child's accidental death spurred the search for another alkaloid, similar to psilocybin, but possibly toxic. Soon, baeocystin was discovered (Leung and Paul 1968; Bennedict et al. 1962). Most researchers are unaware that baeocystin is a misnomer. Baeocystin, by most objective measures, has not been proven to be toxic in the dosages commonly consumed.

Abundant where wood chips and bark (Douglas firs) have been used for landscaping, I often find *P. baeocystis* amongst ivy as it encroaches over the cultivated landscapes near newly constructed buildings. This mushroom is difficult to see against the background of wood chips, as the dark-colored cap is well disguised. In one rhododendron garden I frequent, I find *Psilocybe baeocystis* and *P. cyanofibrillosa* growing side by side. The incurved, sometimes inrolled cap margin and its irregular form are important features. Taxonomically, the closest relative to *P. baeocystis* is *Psilocybe aztecorum,* its subtropical cousin. Both species fade in drying to a light color, often bleaching white from sun exposure. See also *Psilocybe caerulipes, P. caerulescens,* and *P. cyanescens.*

Psilocybe bohemica Sebek

COMMON NAME: the bohemian *Psilocybe*

Cap: 1–4 cm broad. Obtusely conic to campanulate or convex when young, expanding to broadly convex to plane in age. Ochraceous brown to dingy orangish brown, lighter towards the margin, hygrophanous, fading from the disc in drying to straw colored or dingy yellow, eventually a dingy white when dried. Margin even, incurved when young, then decurved. Soon straightening, sometimes irregular or eroded but not undulating with regularity, translucent-striate a third of the way to the center. Surface glabrous when moist, but not viscid and lacking a separable gelatinous pellicle. Flesh whitish to cream colored, soon bruising bluish where injured. **Gills:** Attachment adnate to adnexed, close, light grayish to pallid when young, darkening with spore maturity to yellowish brown, and eventually dingy brown to nearly chocolate brown in age, with margins remaining pallid or whitish

Psilocybe bohemica. **Highly active.**

fringed. **Stem:** 20–60 mm long by 2–5 mm thick. Silky, white overall to dingy brown below, covered with scattered fine fibrils, equal to flexuous, often curved at the apex and near the base, which is typically enlarged and adorned with thick white clusters of rhizomorphs, bruising bluish where injured. Stuffed with a pith, becoming hollow with age, internal flesh white to dingy brown, especially near the base. Partial veil thinly cortinate, white, soon disappearing, not usually fibrillose veil remnants on the stem or cap. **Microscopic features:** Spores grayish purple brown in deposit, 10–14 by 5–7 μ. Basidia 4-spored, rarely 2-spored. Pleurocystidia few, fusiform to lageniform, and similar to cheilocystidia. Cheilocystidia 20–35 by 5–9 μ, lageniform, with a long neck, 20–35 by 5–9 μ. **Habit, habitat, and distribution:** Reported in the autumn from central Europe: the Czech Republic, Austria, and Germany. Probably more widely distributed than presently realized. Lignicolous, fruiting on woody debris around deciduous trees and conifers, including *Betula, Carpinus, Alnus,* and *Piceae.* **Comments:** Potently active. Stijve and Kuyper (1985) found maxima of .80% psilocybin, .02% psilocin, and .03% baeocystin. Gartz published two studies: in 1993, he found maxima of .96% psilocybin, .02% psilocin, and .03% baeocystin, and in 1994, he found 1.34% psilocybin, .02% psilocin, and .03% baeocystin. Gartz (1989) has succeeded in cultivating this mushroom outdoors in wood chips in a garden in Germany, where it persisted for more than four years. It's worth noting that Gartz found complete reproductive barriers between the spores of *P. bohemica, P. cyanescens, and P. azurescens*. Also, the caps of *P. bohemica* become lighter in drying (to a dingy white), in contrast to the two aforementioned species, whose caps become dark in drying (Stamets and Gartz 1995). The method for cultivation is essentially the same that has been outlined for *Psilocybe cyanescens* (Stamets and Chilton 1983) and for *Psilocybe azurescens* (Stamets 1995). First described from the former Czechoslovakia by Sebek (1980), who noted its natural affinities to *Psilocybe collybioides, Psilocybe cyanescens,* and *Psilocybe serbica*. Visually, *P. bohemica* shares many features in common with other ochraceous-colored Psilocybes, *P. cyanofibrillosa* and *P. azurescens.*

Psilocybe brasiliensis Guzman

Cap: 1–3 cm broad. Conic at first, soon convex and eventually campanulate. Surface smooth, viscid when moist, and translucent striate along the cap margin. Reddish brown to brown when moist, hygrophanous, fading to light brown upon drying, becoming bluish in age or where damaged. **Gills:** Attachment adnate, brownish to purplish brown, with whitish margins. **Stem:** 35–80 mm long by 1–4 mm thick, equal to slightly swelling at the base, which is usually adorned with whitish fibrils and/or rhizomorphs. Bruising bluish where injured. **Microscopic features:** Spores dark grayish violet in deposit, subrhomboid to subellipsoid, 5.5–7.7 by 4.2–5.5 μ. Basidia 4-spored. Pleurocystidia 22–29 by 8.8–12 μ, fusoid-ventricose to sublageniform. Cheilocystidia 9.6–13 by 5–7.2 μ, lageniform. **Habit, habitat, and distribution:** Gregarious on grassy soils (*Axonopus compresus*) in a forest of *Araucaria brasiliana* and *Podocarpus*. Known only from Brazil, near Sao Paulo, in March at approximately 1500 meters. Not yet reported outside this type locality. **Comments:** Apparently active, Guzman (1983) first discovered *P. brasiliensis,* and lists it under the psilocybin-

***Psilocybe brasiliensis*.**
Active, potency unknown.

producing varieties because of the bluing reaction. The general shape is reminiscent of some of the temperate Psilocybes like *P. cyanofibrillosa*. Being a grassland species, this mushroom is likely to be widespread and is a good candidate for cultivation in tropical environments. I think if more Brazilians were aware of this species, its range of dominion would be better documented. See also *Psilocybe aztecorum, Psilocybe baeocystis,* and *Psilocybe mexicana*.

Psilocybe caerulea (Kriesel) Noordeloos

= *Stropharia caerulea* Kriesel
= *Stropharia cyanea* (Bolt. ex. Secr.) Tuomikoski

Cap: 3–7.5 cm broad, convex, soon expanding to broadly convex, obtusely umbonate. Color bluish, fading to azure, then yellow to yellowish green, often with the margin remaining tinged bluish green. Viscid when wet from a separable gelatinous pellicle. Fragments of the partial veil sometimes present along cap margin. Cap flesh thin, bluish at first, fading to straw color. **Gills:** Attachment adnate to subdecurrent, close, fawn brown to chocolate brown to dark brown with spore maturity, with paler edges. **Stem:** 40–110 long by 3–12 mm thick, equal to slightly enlarged near the base, bluish green to azure, becoming pale buff. Pruinose above the annulus and covered with fibrillose-floccose scalelike patches below. Partial veil membranous, leaving a fragile, membranous annulus sometimes degrading into an annular zone. **Microscopic features:** Spores light purplish brown in deposit, ellipsoid, 7–9 (10) by

***Psilocybe caerulea*. Activity unknown.**

4.5–5.5 μ. Basidia 4-spored. Cheilocystidia and pleurocystidia 28–55 by 10–16 μ, 2–4 at the apex, clavate-mucronate to fusiform or slightly lageniform with acute apices. **Habit, habitat, and distribution:** A litter mushroom, thriving in garden-like habitats or soils enriched with manure. Widely distributed, reported from the British Isles, Europe, and northwestern North America. **Comments:** Possibly mildly active. No chemical analyses are known to this author. This species differs from its close relative, *Psilocybe aeruginosa,* in its rapidly discoloring cap, the paler gills, and the lack of veil-formed scales on the cap surface, and its generally smaller stature. In Europe, *P. caerulea* is more common than *Psilocybe aeruginosa* (= *Stropharia aeruginosa*). See also *Psilocybe pseudocyanea (= Stropharia pseudocyanea).*

Psilocybe caerulescens Murrill

= *Psilocybe caerulescens* var. *mazatecorum* Heim

COMMON NAME: derrumbes (landslide mushroom)

Cap: 2–9 cm broad. Obtusely campanulate to convex with a decurved margin at first, becoming convex; rarely plane with great age and often having either a small umbo or a slight depression in the center. Margin often bluish, translucent-striate halfway to the center portion of the cap and hanging with fragile whitish veil remnants (appendiculate). Deep olive black in young specimens, strongly hygrophanous, fading with age to a dark reddish brown to chestnut brown near the disc and often darker towards the margins. Margin incurved and often inrolled when young. Surface smooth and slightly viscid to lubricous when moist, pellicle thinly gelatinous but not usually separable. Flesh whitish to dingy brown, moderately thick, and bruising bluish. **Gills:** Attachment sinuate to adnate, close to subclose, and broad. Color grayish to soot brown, with the edges remaining whitish. **Stem:** 40–120 mm long by 2–10 mm thick. Mostly equal but often radicating into a long pseudorhiza. Covered at first with a whitish layer of fibrils, which soon deteriorates—revealing a more sordid brown, smooth surface underneath. Upper regions of the stem characteristically adorned with whitish fibrillose patches. Partial veil cortinate, whitish and copious at first, but soon disappearing. Flesh stuffed and fibrous; bruising bluish, whitish rhizomorphs (bluish when disturbed) present about the base of the stem. **Microscopic features:** Spores dark purplish brown in deposit, subrhomboid to subellipsoid, 6–8 by 4–6 μ. Basidia 4-spored, occasionally

Psilocybe caerulescens. **Moderately to highly active.**

2-spored. Pleurocystidia absent. Cheilocystidia 15–22 by 4.5–6 μ, fusoid with a flexuous neck 1–2.5 μ broad. **Habit, habitat, and distribution:** Gregarious to cespitose, rarely solitary, found in the late spring and summer on disturbed or cultivated grounds often devoid of herbaceous plants. Preferring muddy orangish brown soils. First reported from near Montgomery, Alabama, by Murrill in 1923 on sugarcane mulch, and not redocumented from that locality since. *P. caerulescens* is widespread throughout central regions of Mexico, and also Venezuela and Brazil. **Comments:** A potent species, thirteen pairs of this mushroom were ingested by R. Gordon Wasson during his inaugural session with Maria Sabina, the renowned Mazatec healer. Heim and Hofmann (1958) found .20% psilocybin and 0% psilocin, but from aged specimens. *P. caerulescens* is obviously much more potent than they reported. Along with *Psilocybe aztecorum*, this mushroom is the likely species referred to by Sahagun as the *teonanacatl* used by the Aztecs.

Considerable variation in cap color and the extensive range of this mushroom underscore the probable number of forms this mushroom expresses. The fibrillose patches on the stem seem to be a fairly consistent feature. This mushroom frequently fruits in clusters. Also, very young mushrooms when cut lengthwise show a distinctive inrolled margin in the cross-section of the cap. This shape is reminiscent of the forms the Aztecs portrayed in their art. (See page 13.) This mushroom can be very bitter, otherwise some forms lack any distinctive flavor. See also *Psilocybe zapotecorum* and *Psilocybe weilii* nom. prov.

Psilocybe caerulipes (Peck) Saccardo

COMMON NAME: blue foot

Cap: 1–3.5 cm broad. Obtusely conic, becoming conic-campanulate to broadly convex to plane with age, and may retain a slight umbo. Margin incurved at first, often tinged greenish, very irregular, closely translucent-striate, and decorated at first with fibrillose veil remnants. Cinnamon brown to dingy brown, hygrophanous, fading to pale ochraceous buff. Surface viscid when moist from a gelatinous pellicle, but soon becoming dry and shiny. Flesh thin, pliant, and bluing where bruised. **Gills**: Attachment adnate to sinuate to uncinate, close to crowded, narrow, with edges remaining whitish. Color sordid brown at first, becoming rusty cinnamon. **Stem:** 30–60 mm long by 2–3 mm thick. Equal to slightly enlarged toward the base. White to buff at first, with the lower regions dingy brown at maturity, bluing where bruised. Surface powdered at the apex, and covered with whitish to grayish fibrils downwards. Flesh stuffed with a pith and solid at first but soon becoming tubular. Partial veil thin, cortinate, and forming an evanescent fibrillose annular zone in the superior region of the stem, if at all. **Microscopic features:** Spores dark purplish brown in deposit, ellipsoid, 7–10 by 4–5 μ from 4-spored basidia. Spores from 2-spored basidia are larger. Pleurocystidia absent. Cheilocystidia 18–35 by 4.5–7.5 μ, lageniform, with a thin neck, sometimes forked, 1–2.5 μ broad at apices. **Habit, habitat, and distribution:** Solitary to cespitose on hardwood

Psilocybe caerulipes. Note fine veil remnants near cap margin. Moderately active.

Psilocybe caerulipes. Note change in cap color from drying.

slash and debris, and on or about decaying hardwood logs (particularly birch, beech, and maple), especially along river systems. Growing in the summer to fall after warm rains. Widely distributed east of the Great Plains, throughout the Midwestern and eastern United States. Surprisingly, Guzman made two collections of this species in Mexico (the southern most range of beech [*Fagus* sp.] in North America): the state of Hidalgo and southeastern Zacualtipan. **Comments:** Moderately active; no analyses published. The bluing reaction is variable, more evident in drying, and may take several hours before it can be seen. Although widely distributed, *P. caerulipes* is not found frequently. However, in woodlands where it is known, fruitings tend to persist for years. See also *Psilocybe quebecensis* and *Psilocybe cyanescens.*

Psilocybe coprophila (Bulliard ex Fries) Kummer

= *Psilocybe mutans* Mcknight

Cap: 1–3 cm broad. Convex or hemispheric. Margin sometimes finely appendiculate, usually translucent-striate halfway to disc. Orangish brown to reddish brown. Hygrophanous. Surface viscid when moist from a separable gelatinous pellicle; smooth overall with buff fibrils along the margin. Flesh relatively thin and nearly concolorous with the cap. **Gills:** Attachment adnate, broad, and subdistant. Color grayish brown at first, then deep purple brown to black with spore maturity. **Stem:** 20–60 mm long by 1–3 mm thick. Nearly equal. Yellow to yellowish brown. Surface covered with scattered fibrils and dry. Partial

Psilocybe coprophila. Note color and translucent-striate quality of fresh caps. Not active.

veil thin to absent. Sometimes bruising bluish in the mycelium. **Microscopic features:** Spores dark purplish brown in deposit, subellipsoid, 11–15 by 6.5–9 µ. Basidia 4-spored. Pleurocystidia absent. Cheilocystidia 22–35 (44) by 7.5–8.5 µ. **Habit, habitat, and distribution:** Usually found on cow or horse dung in the spring, summer, and fall. Never cespitose, but scattered to numerous. Widely distributed throughout the temperate to subtropical regions of the world: throughout North America, Central and South America, Europe, Russia, Japan, Australia, New Zealand, Hawaii, and southern Africa. **Comments:** Does not produce psilocybin or psilocin. This petite mushroom, often broader than it is tall, typically grows in groups, well separated. The caps are very hygrophanous and the cap striations—actually gills showing through the translucent flesh—are soon masked by the opaque flesh as the cap dries. Some strains have been reported to be slightly, although weakly, active. Guzman (1983, 227) suggests that analyses by Repke and Leslie may have shown a positive result from a mixed collection. No other reports of activity are known to me. See also *Psilocybe angustispora, Psilocybe subviscida,* and *Psilocybe merdaria.*

Psilocybe crobula (Fries) Singer

= *Geophila crobula* (Fr.) Kuhner and Romagnesi
= *Psilocybe inquilina* var. *crobulus* (Fr.) Holland

Cap: .4–4 cm broad. Convex to broadly convex, expanding to nearly plane with maturity. Dingy brown, hygrophanous, fading to yellowish brown. Surface smooth, translucent-striate, viscid when wet from a separable, gelatinous pellicle. Often adorned with fragile remnants of a veil, soon disappearing. **Gills:** Attachment adnate, clay to dull rust, with whitish edges. **Stem:** 5–12 mm long by 1–4 mm thick, adorned with distinct fibrillose patches, concolorous with cap, more dark reddish brown near the base and lighter above an annular zone. Partial veil cortinate. **Microscopic features:** Spores subellipsoid, thin walled, cigar brown in mass, 6–8 by 3.5–5 μ. Pleurocystidia absent. Cheilocystidia lageniform to sublageniform, 25–45 μ long. **Habit, habitat, and distribution:** Fruiting in the fall on twigs and other wood debris—not on grass. Reported from the northwestern United States, Great Britain, much of Europe, and Russia. Probably more widely distributed. **Comments:** Not known to me as being a psilocybin mushroom. However, Phillips (1981) notes that it is active, without further elaboration or supporting references. Analyses by Høiland (1978) failed to detect any psilocybin or psilocin. *P. crobula* is a small mushroom, which I have always found directly attached to sticks. The fibrillose patches on the stem are quite distinctive, and reminiscent of many small Galerinas. For this reason, I urge caution. See also *Psilocybe inquilina*, a species similar in its macroscopic features and with which it is often confused, *Psilocybe atrobrunnea*, and *Psilocybe washingtonensis*. Guzman and Smith (1978) noted the similarities *Psilocybe crobula* shares with *Psilocybe laticystis* and *Psilocybe subborealis*, two species from the Pacific Northwest.

Psilocybe crobula.

Psilocybe cubensis (Earle) Singer

= *Psilocybe cubensis* var. *caerulescens* (Murr.) Singer and Smith
= *Stropharia cubensis* Earle
= *Stropharia cyanescens* Murr.
= *Stropharia caerulescens* (Pat.) Sing.

COMMON NAMES: golden tops, cubies, san isidros, hongos kentesh

Cap: 1.5–8 cm broad. Conic-campanulate often with an acute umbo at first, becoming convex to broadly convex and finally plane in age with or without an obtuse or acute umbo. Reddish cinnamon brown in young fruiting bodies, becoming lighter with age to more golden brown, fading to pale yellow or white near the margin with the umbo or the center region remaining more darker cinnamon brown. Surface viscid to smooth when moist but soon dry; universal veil leaving spotted remnants on cap but soon becoming smooth overall. Flesh whitish, soon bruising bluish. **Gills:** Attachment adnate to adnexed, soon seceding, close, narrow to slightly enlarged in the center. Pallid to grayish in young fruiting bodies, becoming deep purplish gray to nearly black in maturity, often mottled. **Stem:** 40–150 mm long by 5–15 mm thick. Thickening towards the base in most specimens. Whitish overall but may discolor to yellowish; bruising bluish where injured. Surface smooth to striated at the apex, and dry. Partial veil membranous, leaving a well-developed, white, persistent membranous annulus that often bruises bluish and soon becomes dusted with purplish brown spores. **Microscopic features:** Spores dark purplish brown to violaceous brown in deposit, subellipsoid, 11.5–17 by 8–11 µ. Basidia 2- or 3-spored, but usually 4-spored. Pleurocystidia nearly pear shaped, sometimes mucronate, 18–30 by 6–13 µ. Cheilocystidia fusoid-ventricose with an obtuse or subcapitate apex, sometimes sublageniform, 17–32 by 6–10 µ, with the narrow necks 3–5 µ broad. **Habit, habitat, and distribution:** Scattered to gregarious on dung of bovines (cattle, oxen, yaks, water buffalo), horse, or elephant dung, and on well-manured grounds in the spring, summer, and fall. Found

Young *Psilocybe cubensis*. The tearing of the partial veil creates a membranous ring on the stem.

throughout the southeastern United States, Mexico, Cuba, Central America, northern South America, the subtropical Far East (India, Thailand, Vietnam, and Cambodia), and regions of Australia (Queensland). Typically, the largest fruitings of this species are seen in the two months prior to the hottest period during the year. In the southeastern United States, May and June are the best months for picking, although they can be found up until January. **Comments:** On the psilometric scale of comparative potency, *P. cubensis* gets a rating of "moderately potent," with maxima reported by Heim and Hofmann (1958) of .50% psilocybin and .25% psilocin, while Gartz (1994) reported .63% psilocybin and .11% psilocin. Stijve and de Meijer (1993) found .15% psilocybin and .50% psilocin in a Mexican strain, and .15% psilocybin and .33% psilocin in an Amazonian strain. Analyses of *P. cubensis* vary substantially due to a series of complex variables. Bigwood and Beug (1982a) found a fourfold variation in potency in cultivated specimens and up to a tenfold variation in potency from wild specimens. (In one collection, Beug and Bigwood found an extraordinary 1.3% psilocybin and .35% psilocin!) Mushrooms grown indoors seem consistently more potent than field-collected specimens, probably due to nutritional factors (precursors) and protection from the damaging effects of ultraviolet radiation. Gartz (1989) determined that psilocin levels of flushes were naturally low (.1%) from a sterilized mixture of cow dung and rice (2:1), but could be raised up to 3.3% with the addition of only 25 mil-

***Psilocybe cubensis*, Palenque, Mexico. Note bruising on stem and on the partial veil, which becomes a membranous ring. Moderately to potently active.**

***Psilocybe cubensis* fruiting from elephant dung in India.**

ligrams of tryptamine into 10 grams of substrate. Furthermore, his study showed that at least 22% of psilocybin was derived from the introduced radioactively tagged tryptamine. This study reinforces the concept that substrate composition affects potency.

This species, the most majestic of the Psilocybes, is easy to recognize by its size, golden color, the well-formed membranous annulus, the blue-staining stem and veil, and the coprophilic habitat. *P. cubensis* is thought to have come with the Spaniards during the Cortés and subsequent missionary expeditions. Although it is not known from Spain, the trade routes at that time could have carried spore mass from subtropical African islands to the New World. This is the only species carrying a reference to Spanish Catholicism: "san isidro" is an epithet used by the indigenous peoples (Guzman 1983). Although *P. cubensis* is widely sold to tourists, the shamans of Oaxaca prefer to use *Psilocybe caerulescens, Psilocybe aztecorum, Psilocybe zapotecorum,* or *Psilocybe mexicana.*

P. cubensis has been widely cultivated in the United States and Europe since the publication of Oss and Oeric (1976), Stamets and Chilton (1983), and other books that showed techniques for home cultivation. Most of the spores were brought back by travelers to Mexico, Guatemala, Ecuador, Colombia, and the Amazon in the mid seventies. Hence, strains carried the names of their origins, such as Amazonian, Palenque, Matias Romero, and Ecuadorian. These strains were widely distributed throughout the world, and in time had a ripple effect that increased their availability, and hence their popularity. *Psilocybe subcubensis* Guzman is virtually identical, differing only in the smaller size of its spores, which are (9.9) 11–13 (14) by 6.1–7.1 μ, and slightly smaller pleurocystidia. *Psilocybe subcubensis* has been collected throughout much of subtropical Mexico, Colombia, Bolivia, Ecuador, Honduras, El Salvador, Venezuela, and Australia. Probably more widely distributed. The reports of *P. subcubensis* (Stamets 1978; Ott and Bigwood 1978) from California are probably from outdoor cultivated specimens.

Psilocybe cyanescens Wakefield

COMMON NAMES: cyans, blue halos, wavy-capped *Psilocybe*

Cap: 2–4 (5) cm broad. Obtusely conic to conic-convex at first, usually soon expanding to broadly convex to nearly plane in age with an undulating or wavy margin. Margin translucent-striate. Chestnut brown in young specimens, becoming more caramel colored with age, hygrophanous, fading to dark yellowish brown or ochraceous in drying. Surface smooth and viscid when moist from a sometimes separable gelatinous pellicle. Context nearly concolorous with the cap and bruising bluish. **Gills:** Attachment adnate to subdecurrent, close to subdistant, broad. Color is cinnamon brown, becoming deep smoky brown with the edges renaming paler. **Stem:** 20–80 mm long by 2.5–5 mm thick. Often curved and somewhat enlarged at the base, stiff but not pliant. Whitish overall, readily bruising bluish. Surface silky, covered with fine fibrils and often with long whitish rhizomorphs protruding about base of stem. Partial veil copiously cortinate, snow-white, rapidly deteriorating to an obscure annular zone, if at all. **Microscopic features:** Spores dark purplish brown in deposit, elongate-ellipsoid, 9–12 by 5–8 μ. Basidia 4-spored. Pleurocystidia not reported by Wakefield, but collections from the Pacific Northwest have abundant, capitate pleurocystidia 17–33 by 5–8.8 μ, fusoid-ventricose to subpyriform, sometimes mucronate, more common near the gill edge. Cheilocystidia (12) 16–27 (30) by (5) 6.6–8.8 μ, sublageniform to fusoid-ventricose, cylindrical at base, with an extended single or split neck 6 by 1.5–3.5 μ. **Habit, habitat, and distribution:** Scattered to gregarious in humus enriched with woody debris,

Psilocybe cyanescens. **Note wavy cap margins. Moderately to highly active.**

amongst leaves and twigs, in wood chips, sawdust, or in debris fields rich with rotting wood. Often under mixed woods at the edges of lawns, along paths, and in heavily mulched rhododendron and rose gardens. Found in the fall to early winter in the Pacific Northwest. Reported from the western coastal regions between San Francisco, California, to southern Alaska, and also widely spread throughout the United Kingdom and across much of temperate Europe, from Italy, Germany, Spain, and Sweden. **Comments:** Moderately to highly potent. Beug and Bigwood (1982b) reported maxima of 1.68% psilocybin and .28% psilocin. Gartz (1994) reported .30% psilocybin, .51% psilocin, and .02% baeocystin. Stijve and Kuyper (1985) found a maximum of .85% psilocybin, .36% psilocin, and .03% baeocystin. The wavy cap margin, the color of the cap, and the copious nature of the partial veil distinguish this species. However, the original descriptions of *P. cyanescens* consistently note that pleurocystidia are absent, or if present are only near to the gill edge (Wakefield 1946; Singer and Smith 1958b; Guzman 1983). My own studies of *P. cyanescens* with the scanning electron microscope reveal abundant, capitate pleurocystidia. If one accepts the presence of pleurocystidia as being taxonomically significant, the description of this species needs to be emended, or a new taxon encompassing these features described. I know of several rhododendron gardens faithfully maintained by dedicated caretakers who have unwittingly succeeded in promoting perennial patches of *P. cyanescens.* This species loves wood-chip trails that meander through gardenlike settings bordered by rhododendrons and shade-providing shrubs. *P. cyanescens* grows well with lupines, azaleas, and other bushes associated with the coastal, temperate planes. Andrew Weil (1975; 1977) brought this mushroom to the forefront of awareness after he collected this species in Oregon. *P. cyanescens* is macroscopically similar to *P. azurescens,* and differs in the form of the cap margin, and the lack of distinct umbo. See also *Psilocybe serbica, Psilocybe bohemica, Psilocybe subaeruginosa, Psilocybe cyanofibrillosa,* and *Psilocybe mairei.*

A sacred *Psilocybe cyanescens* patch.

Psilocybe cyanofibrillosa Stamets and Guzman

= *Psilocybe rhododendronensis* Stamets nom. prov.

COMMON NAMES: rhododendron *Psilocybe,* blue-haired *Psilocybe*

Cap: 1.4–3.5 cm broad. Conic to convex to broadly convex, eventually plane in age, typically not umbonate. Color deep chestnut brown, hygrophanous, fading in drying to pale tan to yellowish brown, even dingy grayish white in drying. Surface viscid when moist from a separable gelatinous pellicle. **Gills:** Attachment adnate to adnexed, to slightly subdecurrent in age, light grayish when young, becoming purplish brown with maturity with whitish edges. **Stem:** 30–70 mm long by 2–4 mm thick, straight to flexuous, equal to enlarged near the base, longitudinally striate, and adorned with fine fibrils that become bluish when handled. Yellow brown to light tan underneath. Partial veil white, cortinate, copious, leaving fibrillose veil remnants, sometimes a fragile annular zone on the upper regions. Flesh brownish, bruising bluish. **Microscopic features:** Spores purplish brown in deposit, subellipsoid, (9) 9.5–11 (12) by (5.5) 6–6.6 (7) μ. Basidia 4-, rarely 2-spored. Pleurocystidia absent. Cheilocystidia fusiform to lanceolate, 22–33 by 5.5–7 μ, with an elongated, forking neck, 1–1.5 μ thick at apex. **Habit, habitat, and distribution:** Growing gregariously to scattered primarily along the coastal regions from Northern California (Eureka/Arcata) north to British Columbia. Associated with bush lupines and especially common on flood plains on river estuaries

***Psilocybe cyanofibrillosa* in its adolescent stage. Weakly to moderately active.**

Psilocybe cyanofibrillosa in its adult stage.

flowing into the Pacific ocean. Also frequently found in coastal rhododendron gardens and nurseries. **Comments:** Weakly to mildly active, containing up to .21% psilocybin and .062% psilocin (Beug and Bigwood in Stamets et al. 1980). Although this species has a strong bluing reaction, due to its psilocin content, it loses much of its potency from handling and in drying. Actual potency of fresh specimens is probably much higher than chemical analyses revealed. To date, *Psilocybe cyanofibrillosa* has only been reported from the Pacific Coast region of North America. See also *Psilocybe azurescens, Psilocybe caerulipes, Psilocybe cyanescens,* and allies.

Psilocybe eucalypta Guzman and Watling

COMMON NAME: eucalytpus *Psilocybe*

Cap: 1.5–3.8 cm broad. Convex, expanding to broadly convex to nearly plane, often with a central shallow umbo. Surface smooth to slightly translucent-striate near the margin, viscid when moist, reddish brown when young, becoming ochraceous, hygrophanous, fading in drying to dull straw colored. Flesh whitish, bruising bluish. **Gills:** Attachment adnate, dull brown, soon becoming purplish brown, and eventually dark violaceous gray with whitish edges. **Stem:** 65–86 mm long by 2–2.5 mm thick. Covered with silky fibrils, equal to swollen at the base (4–5 mm thick) to which whitish rhizomorphs are attached. Partial veil cortinate, leaving a fugacious fibrillose annular zone in the upper regions. **Microscopic features:** Spores purplish brown in deposit, subellipsoid, 9–13 by 5.5–6.6 µ. Basidia 4-spored.

Pleurocystidia 17–30 by 5.5–7.7 µ, fusoid-ventricose to mucronate to sublageniform with short necks 2–3 mm thick. Cheilocystidia 15–25 by 4.4–6.6 µ, fusoid-ventricose to sublageniform, with elongated necks 4–5 µ long by 3–4 µ broad. **Habit, habitat, and distribution:** Solitary to gregarious on soils rich in woody debris, often in grassy areas with *Eucalyptus* trees. Reported from New South Wales, eastern Australia. **Comments:** Thought to be moderately active, but no analyses have yet been reported. Chang and Mills (1992) sought to show synonymy between *Psilocybe australiana, Psilocybe eucalypta, Psilocybe subaeruginosa,* and *Psilocybe tasmaniana,* although, in the opinions of Guzman, Bandala, and King (1993) and myself, they failed to conclusively prove their case. (See Comments under *Psilocybe subaeruginosa,* on page 153.) Additionally, I find it difficult to accept their concept that, given the nature of close relatives in the genus *Psilocybe,* the same species can be both a coprophilic and lignicolous species. Hence, I am following Guzman and treating these taxa as separate entities until further studies prove otherwise.

Psilocybe eucalypta. **Probably moderately active.**

Psilocybe fimetaria. **Moderately active.**

Psilocybe fimetaria (Orton) Watling

= *Psilocybe fimetaria* (Orton) Singer
= *Psilocybe caesioannulata* Singer
= *Stropharia fimetaria* Orton

Cap: (.5) 1–2.5 (3.6) cm broad. Conic to convex, eventually subcampanulate, expanding to broadly convex, and typically acutely umbonate with a sharp papilla. Surface smooth to translucent-striate near the margin, viscid when moist from a thick, separable gelatinous pellicle. Color pale reddish brown to honey to ochraceous, hygrophanous, fading in drying to yellowish olive to ochraceous or yellowish buff. Flesh whitish to honey colored, bruising bluish where injured. **Gills:** Attachment adnate, sometimes sinuate or uncinate, whitish clay at first, eventually dark purplish brown at maturity, with whitish edges. **Stem:** (20) 40–65 (90) mm long by (.5) 1–3 (4) mm thick. Equal to slightly swollen at the base. Color whitish at first, soon reddish brown or honey colored, and sometimes with grayish bluish green tones. Surface covered with whitish fibrillose patches to a fairly persistent, superior densely fibrillose to membranous annulus that develops from a thickly cortinate partial veil. **Microscopic features:** Spores dark purplish brown in deposit, subellipsoid or ellipsoid, (9.5) 11–14 (16) by 6.5–8.5 (9.5) μ. Basidia 4-spored. Pleurocystidia absent. Cheilocystidia 20–32 by 4–8 μ, ventricose-fusiform or lageniform with a narrow neck, often flexuous, 4–15 by .5–1.5 μ, occasionally branched. **Habit, habitat, and distribution:** Grows solitary to gregariously on horse manure, in grassy areas or in rich soils, and often fruits in large rings. Known from Canada (British Columbia and New Brunswick), the Pacific Northwest (Washington, Oregon, and Idaho), Chile, Great Britain, and Europe (Finland, Norway, and the Czech Republic). Probably more widely distributed than presently reported. Generally found in October and November, but in Chile, this mushroom has been collected in August. Its fruiting range is probably much greater than presently reported. **Comments:** Moderately active. At first glance, this species resembles other annulate Psilocybes such as *P. subaeruginascens* or *P. stuntzii*, but *P. fimetaria* favors dung. It also bears similarity to *Psilocybe semilanceata*, except for the superior, persistent, and often membranous annulus that can be bluish toned. *Psilocybe subfimetaria* is closely related, except that it tends not to be sharply papillate and has smaller spores (10–12 by 6–7.7 µ). See also *Psilocybe liniformans* and *Psilocybe stuntzii* var. *tenuis*.

Psilocybe herrerae Guzman

Cap: .8–1 (1.9) cm broad. Conic to convex, sometimes papillate. Surface smooth to dry, finely fibrillose-rimose. Brown to sepia brown, hygrophanous, fading in drying to yellowish, sometimes with bluish tones. Flesh white, bruising bluish where injured. **Gills:** Attachment adnate, becoming purplish brown to brownish violet at spore maturity. **Stem:** 30–80 (110) mm long by 1–2.5 mm thick. Equal overall, slightly narrowing near the apex and tapering into a long pseudorhiza. Brownish or nearly the same color as the cap, quickly bruising bluish, especially from the base upwards. Surface covered with fragile fibrillose-floccose patches. **Microscopic features:** Spores dark purplish brown in deposit, subellipsoid in side view and subrhomboid in face view, 5–6 by 3.3–4.9 µ. Basidia 4-spored. Pleurocystidia 12–25 by 6–8.8 µ, primarily fusoid-ventricose, few are ventricose-rostrate with an abbreviated neck. Cheilocystidia 13–23 (27) by 5.5–7.7 µ, variable in form, 7–11 by 1.5–2.5 µ, sometimes

***Psilocybe herrerae*. Note long pseudorhiza. Moderately active.**

irregularly forking. **Habit, habitat, and distribution:** Solitary to gregarious, often along road cuts, in soils high in sand and clay, and in open forests dominated by pines, sweetgums, and oaks. Found in June and July in Chiapas and Veracruz, Mexico. **Comments:** Judging by the rapid bluing reaction, *P. herrerae* is probably potent, but it has not yet been analyzed. The long pseudorhiza delineates this species from most others. See also *Psilocybe wassoniorum*, macroscopically similar (except for the length of the pseudorhiza) but lacking pleurocystidia.

Psilocybe hoogshagenii Heim sensu lato

= *Psilocybe caerulipes* var. *gastonii* Singer
= *Psilocybe zapotecorum* Heim sensu Singer
= *Psilocybe semperviva* Heim and Callieux

COMMON NAME: *pajaritos de monte* (little birds of the woods)

Cap: (.7) 1–2.5 (3) cm broad. Conic to campanulate to convex with an acute, extended papilla (up to 4 mm long). Surface slightly viscid when wet, smooth, often ridged halfway to the disc. Reddish brown to orangish brown to yellowish, hygrophanous, fading in drying to straw colored, and bruising blue or blue black. **Gills:** Attachment adnate to adnexed, pale brown to coffee colored, and eventually purplish black at maturity. **Stem:** (30) 50–90 (110) mm long by 1–3 mm thick. Equal to slightly thickened near the base, flexuous, sometimes twisted. Whitish to brownish red near the base, easily bruising bluish

Psilocybe hoogshagenii. **Moderately to highly active**

to bluish black. Partial veil thinly cortinate, fragile, soon disappearing **Microscopic features:** Spores dark purplish brown in deposit, rhomboid to subrhomboid, (5) 6.5–8 (9.6) by 4–5.6 μ. Basidia 4-spored, rarely 2-spored. Pleurocystidia 16–36 by 8–12 μ, ventricose to clavate, often irregular. Cheilocystidia (15) 19–35 by 4.4–6.6 μ, lageniform, narrowing into a long neck 1–3 μ, either acute or subcapitate at the apex. **Habit, habitat, and distribution:** Solitary to gregarious in muddy clay soils in subtropical coffee plantations. Found in June and July in Mexico (Puebla, Oaxaca, and Chiapas) and in February in Argentina. Also reported from Brazil and Colombia. **Comments:** Moderately active. Specimens from Brazil yielded up to .30% psilocybin and .30% psilocin (Stijve and de Meijer 1993). A variety of this mushroom, *Psilocybe hoogshagenii* Heim var. *convexa* Guzman is only slightly umbonate, has a convex cap, and is conspecific with *Psilocybe semperviva* Heim and Callieux. This variety is most common in the state of Puebla, Mexico, and to a lesser degree in the states of Oaxaca and Hidalgo, fruiting from June to August. One of the most unusual-looking Psilocybes yet discovered, this mushroom is quite potent. Heim and Hofmann (1958) found .6% psilocybin and .10% psilocin (as *Psilocybe semperviva* Heim and Caillieux) from cultivated specimens. Guzman (1983) reported that this mushroom grows at 1000–1800 meters in elevation (Argentina) and is commonly seen by coffee growers who report massive flushes coming up in unison and soon disappearing. See also *Psilocybe brasiliensis.*

Psilocybe inquilina (Fries ex Fries) Bresadola

= *Psilocybe ecbola* (Fries) Singer

COMMON NAME: grass-rotting *Psilocybe*

Cap: (.5) 1–2 cm broad. Hemispheric to convex, expanding to broadly convex to nearly plane to slightly umbonate. Reddish brown to brick brown to tan or yellowish brown, hygrophanous, fading in color to straw. Margin translucent-striate most way to the disc when moist from a separable gelatinous pellicle, becoming opaque in drying, and sometimes decorated with whitish floccose scales. **Gills:** Attachment adnate to subdecurrent, reddish brown to purplish gray-brown, with concolorous edges. **Stem:** 20–40 mm long by 1.5–2 mm thick. Equal, flexuous, hollow, whitish to reddish brown, adorned with whitish to

***Psilocybe inquilina* is typically attached to matted grass. Not active.**

brown fibrils toward the base, which often has whitish mycelium attached. Partial veil cortinate, soon disappearing. **Microscopic features:** Spores purplish brown in deposit, 7–8.8 (10) by 4.5–6.6 µ, subrhomboid to subellipsoid in face view, subellipsoid in side view. Basidia 4-spored. Pleurocystidia absent. Cheilocystidia lageniform to sublageniform, (15) 18–38 by 5–8 µ, with a long neck 2.5–3.8 µ broad. **Habit, habitat, and distribution:** Solitary to gregarious, commonly on the base of grass stems in open areas, occasionally on rotting twigs or in rich soils. A temperate species, *P. inquilina* is widely distributed, reported from North America (California, Oregon, and Washington), South America (Argentina, Chile, and Uruguay), and Europe (Denmark, Finland, France, Hungary, Sweden, and Switzerland). Undoubtedly more widely distributed than presently reported. **Comments:** Not known to be active. I frequently find *P. inquilina* growing from the bases of matted, rotting grass in the fall. The mushrooms are difficult to harvest without either the stems breaking or clumps of dead grass being pulled up with each specimen. This feature, combined with the nearly decurrent gills, the convex cap, and the deeply translucent-striate margins, distinguishes this species macroscopically. See also *Psilocybe crobula* and *Psilocybe montana*.

Psilocybe kashmeriensis Abraham

Cap: 3–4 cm broad. Campanulate at first, soon expanding to convex with a distinct acute umbo, ochraceous to chrome yellow, fading in drying to pale brown. Surface smooth, viscid when moist, soon drying and often cracking, appendiculate with whitish remnants of the veil adhering to the margin, which is incurved when young and often elevated in age. **Gills:** Attachment adnate to adnate-sinuate, distant, dull yellow, darkening with age, becoming dark purplish brown at maturity and often mottled. **Stem:** 40–80 mm long by 6–8 mm thick, even, and swelling towards the base which radiates clusters of white rhizomorphs. Partial veil cortinate, leaving an apical fibrillose zone darkened with spores. **Microscopic features:** Spores dark purplish brown in deposit, ellipsoid, 8–10.5 by 5–7 μ. Basidia 4-spored. Pleurocystidia common, similar to cheilocystidia. Cheilocystidia fusoid to clavate to capitate, with an obtuse apex, sometimes forked, 20–50 (60) by 5–8 μ. **Habit, habitat, and distribution:** Grows gregariously, saprophytizing camel grass or lemon grass (= *Cymbopogon jawarancus*), an economically significant resource for various oils, including citronella. This species grows on the cut aerial clumps of lemon grass and is especially prolific the second and third year after planting. First reported from Kashmir, India, by Abraham (1995). **Comments:** Closely related to *Psilocybe squamosa*. The activity of *P. kashmeriensis* has not been documented. Its affection for tall grass is shared by many other Psilocybes. Although not bruising bluish, this mushroom may be active, as several other grass-loving species such as *Psilocybe semilanceata* and *Psilocybe strictipes* often show no bluing reaction even though they can be quite potent, due to their low psilocin content.

Psilocybe liniformans Guzman and Bas

COMMON NAME: blunted-grassland *Psilocybe*

Cap: 1–2.5 cm broad. Convex to broadly convex, sometimes broadly umbonate, but not papillate. Surface smooth, viscid when moist from a separable gelatinous pellicle. Dull grayish ochraceous brown, or slightly olivaceous, hygrophanous, fading in drying from the center, becoming straw brown, sometimes bluish green in tone. **Gills:** Attachment adnexed, close to distant, dark chocolate brown to purplish brown, with gelatinous and removable elastic edge. **Stem:** 14–30 mm long by 1–2 mm

***Psilocybe liniformans* var. *americana*. Weakly to highly active.**

thick. Equal to swelled towards the base. Whitish to pale brownish, darker below, bruising bluish where injured, especially near the base and the apex. Surface pruinose above, and finely fibrillose in the lower regions. Partial veil thin, soon disappearing. **Microscopic features:** Spores dark grayish–purple brown in deposit, ellipsoid in both side and face views, 12–14.5 by 7.5–8.8 (10) µ. Basidia 4-spored. Pleurocystidia absent. Cheilocystidia 22–33 by 5.5–9 µ, lageniform with an extended neck more than 6 µ long by 1.5–2.5 µ thick. **Habit, habitat, and distribution:** Scattered to gregarious on horse dung or in manure-enriched soil in meadows and pastures. Grows in late summer through autumn. The European form is known only from the Netherlands. *P. liniformans* var. *americana* has been collected in Washington, Oregon, and Michigan. This same variety has also been reported from Chile (Guzman 1983). **Comments:** Weakly to moderately active. Stijve and Kuyper (1985) found maxima in the European variety of .16% psilocybin, no psilocin, and .005% baeocystin. Beug and Bigwood reported .59–.89% psilocybin and 0% psilocin in the American form (Stamets et al. 1980). However, both analyses were based on single collections. *P. liniformans* var. *liniformans* Guzman and Bas is common in Europe. Typically, this mushroom grows on dung and has a gill edge that is thinly separable. The American variety, *P. liniformans* var. *americana* Stamets and Guzman, grows in rich pastures or in grasslands, and has a gill edge that cannot be removed, but is otherwise similar. See also *Psilocybe strictipes* and *Psilocybe semilanceata.*

Psilocybe luteonitens (Peck) Saccardo

= *Stropharia umbonatescens* (Peck) Saccardo

COMMON NAMES: the umbonate, dung-dwelling *Stropharia*

Cap: 1–4 cm broad. Conic-campanulate with a distinct umbo, expanding to convex to broadly convex to nearly plane with or without an acute-to-low umbo. Margin decorated with fragile whitish remnants of the partial veil. Yellowish to pale ochraceous brown toward the disc. Surface viscid when moist from a separable gelatinous pellicle. Flesh thin, pallid. **Gills:** Attachment adnate to subdecurrent, close, broad. Color whitish at first, then grayish, and eventually purplish brown. **Stem:** 35–70 mm long by 2–4 mm thick. Equal and slender. Pallid to yellowish, lighter than the cap. Surface powdered above the annular region, and initially covered with fine fibrils below, but often smooth with age. Partial veil thinly membranous, floccose, leaving an obscure evanescent annular zone of fibrils, if at all. **Microscopic features:** Spores dark purplish brown in deposit, subellipsoid to ellipsoid, 15–19 (22) by 10–11 μ. Basidia 2-spored. Pleurocystidia absent. Cheilocystidia 25–45 by 3–7 μ, cylindrical to sublageniform with a flexuous, elongated apex. **Habit, habitat, and distribution:** Gregarious on dung in the fall in the Pacific Northwest. Reported from Washington, Oregon, Idaho, Michigan, and New York in the summer to early fall. Also reported from Mexico, Europe, and Asia. **Comments:** The habitat preference, distinctly umbonate and yellowish cap, and the white membranous partial veil all make this species easy to recognize. Additionally, the veil remnants usually adhere to the cap margin rather than the stem, especially evident when young. See also *Psilocybe semiglobata* and *Psilocybe fimetaria.*

Psilocybe luteonitens. **Activity unknown.**

Psilocybe magnivelaris (Peck apud Harriman) Noordeloos
= *Psilocybe percevalii* (Berkeley and Broome) Orton
= *Stropharia percevalii* (Berkeley and Broome) Saccardo
= *Stropharia magnivelaris* Peck apud Harriman

Cap: 1.5–6 cm broad. Obtusely umbonate to campanulate to convex, expanding to broadly convex to plane and often umbonate, with an elevated margin in age. Surface viscid when moist from a thin gelatinous pellicle, smooth, covered with scattered, white, floccose scales, increasing towards margin. Pale grayish yellow to ochraceous to brownish orange, and darker towards the disc, not strongly hygrophanous. **Gills:** Attachment adnate to sinuate, broad, close, pallid white at first, soon grayish brown and finally dark purplish brown with whitish fringed margins. **Stem:** 50–85 mm long by 4–7 mm thick. Hollow, equal to enlarged towards the apex and tapering below. White to dingy yellowish. Partial veil membranous, leaving a thick, white membranous annulus, often flaring, which can deteriorate into an annular zone, below which the surface can be covered with fibrillose patches. Flesh moderately thick, firm, whitish, and not bruising. **Microscopic features:** Spores dark purplish brown in deposit, smooth, ellipsoid 13–15 by 6–8 µ. Basidia 4-spored. Pleurocystidia absent. Cheilocystidia 33–44 by 4–5 µ, nearly clavate to sublageniform, with an elongated neck 3–4 µ thick. **Habit, habitat, and distribution:** Scattered to gregarious, preferring sandy soils, alluvial plains, and/or soils rich in woody debris of *Salix* (willow) and *Alnus* (alder). Found from May to November in the United States (Alaska, Washington, Oregon, possibly Colorado), northern Europe, and the British Isles. Likely to be much more widely distributed than presently reported. **Comments:** Not active, edibility unknown. The well-developed membranous annulus, the nonbluing flesh, its modest size, and its habitat, all give clues to its identification. This mushroom is similar to *Psilocybe squamosa* and *Psilocybe thrausta,* and can be separated with certainty by the length of the cheilocystidia. *Stropharia riparia* is generally similar. See also *Psilocybe subaeruginascens.*

Psilocybe mairei Singer

= *Psilocybe maire* Singer sensu Guzman

= *Hypholoma cyanescens* Maire

Cap: 1.5–3.5 cm broad. Convex to campanulate to conic-campanulate, expanding with age, but not umbonate. Surface viscid when moist from a separable gelatinous pellicle. Orangish brown, becoming olive toned, hygrophanous, fading in drying to yellowish white. Flesh amber, bruising bluish where injured. **Gills:** Attachment adnate, pallid at first, soon darkening, becoming purplish brown with the edges remaining whitish fringed. **Stem:** 25–75 mm long by 2–5 mm thick. Equal to slightly enlarged towards the base. Whitish to yellowish white, bruising bluish where injured. Surface pruinose above and finely fibrillose below. Stuffed with a silky pith. Base of stem adorned with thick, radiating white rhizomorphs. Partial veil cortinate, white, leaving fibrillose remnants on the stem. **Microscopic features:** Spores dark purplish brown in deposit, elongate-ellipsoid, 10–12 (13.5) by 5.5–7 µ. Basidia 4-spored. Pleurocystidia absent or near to gill edge. Cheilocystidia 30–40 by 6–8 µ. Variable in form: lageniform, fusiform, or ampullaceous. **Habit, habitat, and distribution:** Known only from North Africa (Morocco and Algeria) in October through December. Grows gregariously on soil rich in woody debris, in forests mixed with pine (*Pinus pinaster*), fir (*Abies pinsapo*), and oak (*Quercus ilicis* and *Quercus pyrenaica*). The identification of collections from Europe as *P. mairei* are doubtful, and

LEFT: *Psilocybe mairei.* RIGHT: *Psilocybe mairei* growing from seed cone of *Pinus pinaster.* Estimated to be moderately active.

according to Guzman (1983) were probably *Psilocybe serbica* and allies. **Comments:** Probably potent, judging by the bluing reaction, although no analyses have been published. This is the only wood-decomposing, bluing *Psilocybe* reported from North Africa. Prior to the expansion of the Sahara desert, North Africa enjoyed a moister climate and undoubtedly hosted many more mushrooms than are known today. One wonders if *P. mairei* is a surviving remnant of a species that was once much more common. I am reminded of the cave art representing a beelike shaman undergoing a mushroom experience from the Tassili plateau in Southern Algeria, and I'm struck by the fact that this is the only species of *Psilocybe* presently known in that area. (See page 17.) Given that the Tassili plateau was running with rivers at the time the artist lived, the alluvial plains would have been perfect for supporting species like *P. mairei.* Could this have been the mushroom the artist used? See also *Psilocybe cyanescens*, *Psilocybe serbica*, and allies.

Psilocybe makarorae Johnston and Buchanan

Cap: 1.5–3.5 cm in diameter. Conic to campanulate, expanding in age to broadly convex, usually with a pronounced umbo. Surface tacky to dry. Yellowish brown to orangish brown, fading in drying, and lighter towards the margin, which is striate when moist. Flesh whitish, bruising greenish blue where injured. **Gills:** Attachment adnexed, pale grayish brown with concolorous edges. **Stem:** 30–60 mm long by 2–4 mm thick, equal, white to brownish near the base, which radiates white rhizomorphs. Partial veil cortinate, leaving fibrillose veil remnants along the cap margin when young, soon disappearing, but not forming an annular ring on the stem. Surface finely fibrillose and silky. **Microscopic features:** Spores dark purplish brown in deposit, ellipsoid in side view, ovoid to subrhomboid in face view, (6.5) 7.5–10 by 4.5–6.5 μ. Basidia 4-spored. Pleurocystidia similar to cheilocystidia, ventricose-rostrate to mucronate, 4–8 μ thick, with a simple neck, 2.5–4 μ long. Cheilocystidia 18–26 by 6–9 μ, ventricose-rostrate to mucronate with a short, simple neck 3–5 μ long. **Habit, habitat, and distribution:** Found scattered to gregarious on rotting wood and twigs (*Nothofagus*) near lakes and picnic grounds in the vicinity of Makarora, New Zealand (Otago Lakes and Franz Josef Glacier), in the fall. Probably more widely distributed. **Comments:** Potency unknown but probably moderately active given the bluing reaction. This newly described species by Johnston and Buchanan (1996) resembles *Psilocybe caerulipes*, but the presence of pleurocystidia and longer-necked cheilocystidia delineates *P. makarorae* from this taxon. Worthy of note is that the two authors work closely with, are consulted by, and paid by law-enforcement officials to help in the prosecution of unlucky collectors. See also *Psilocybe australiana, Psilocybe eucalypta,* and *Psilocybe subaeruginosa.*

Psilocybe mammillata (Murrill) Smith

Cap: (.5) 1–3 cm broad. Conic, expanding to campanulate at maturity, and often umbonate. When moist, surface smooth, slightly striate near the margin, which is often adorned with minute remnants of the partial veil. Reddish brown to brownish olivaceous, hygrophanous, fading to beige or dirty yellowish orange from the disc, occasionally with blackish tones. Flesh pale brown, bruising bluish. **Gills:** Attachment adnexed, pale brown to dark purplish brown, thin, with whitish edges. **Stem:** 15–30 mm broad by 1–2 mm thick. Equal to slightly swelling at the base from which white mycelium radiates. Yellowish at first, soon reddish brown with maturity, often retaining whitish patches near the base. Flesh reddish brown, bruising bluish where injured. Partial veil cortinate, fragile, soon disappearing in age. **Microscopic features:** Spores dark purplish brown in deposit, subellipsoid in side view, rhomboid to subrhomboid in face view, 5–6.5 (8) by 3.5–5.5 μ. Basidia 4-spored. Pleurocystidia absent or only near to the gill edge. Cheilocystidia 12–17 by 4.5–5 μ, fusoid-ventricose with elongated necks, 3.3–5 by 1–2.5 μ. **Habit, habitat, and distribution:** Solitary to gregarious, infrequently cespitose, in soils rich in woody debris, in humus, and sometimes on clay soils. Found along trails, shady banks, and in coffee plantations. Discovered by Dr. Thiers in Florida (Highlands Hammock State Park). Also reported from Jamaica, Mexico, and Bolivia. **Comments:** Guzman et al. (1993) reports that this species is active, although no analyses are cited. The classic bluing reaction leaves little doubt about its activity, although estimation of potency would be purely speculative. This mushroom is probably widely distributed throughout much of Florida but goes unrecognized by most hunters of the more massive *Psilocybe cubensis.*

Psilocybe merdaria (Fries) Ricken

Cap: 1–4 cm broad. Campanulate-hemispheric to convex to broadly convex, and sometimes slightly umbonate, finally expanding to plane with age. Margin translucent-striate when moist, and often appendiculate with remnants of the thin partial veil. Cinnamon brown to livid brown when moist, fading to ochraceous or yellowish brown, and remaining darker at the disc. Surface smooth and only moist to subviscid when wet. **Gills:** Attachment adnate to subdecurrent, close to broad. Yellowish at first, darkening with spore maturity to a dark brown. **Stem:** 20–40 mm long by 1–3 mm thick. Pale yellowish to pallid. Surface covered by fine, dry fibrils. Flesh stuffed with a fibrous pith, tough, but soon becoming hollow. Partial veil thinly membranous, fugacious, soon deteriorating to an annular zone of fibrils in the median to lower regions of the stem, usually darkened by spores. **Microscopic features:** Spores dark purple-brown in deposit, subellipsoid, 10–14 by 7–9 µ. Basidia 4-spored. Pleurocystidia absent. Cheilocystidia 20–33 by 6.6–8.8 µ, fusoid-ventricose to sublageniform with a short neck 3.3–4.4 µ thick. **Habit, habitat, and distribution:** Scattered to numerous on dung. Reported from California, Oregon, Washington, and the northern Midwest of the United States. Widely spread throughout the world, this mushroom has been collected in Canada, Europe, the former USSR, and Japan. *P. merdaria* is probably more widespread than the literature presently indicates. It prefers a temperate zone. **Comments:** Not known to be active; not sufficiently studied for

Psilocybe merdaria. **Not known to be active.**

chemical content. In temperate zones, *P. merdaria* is a common dung dweller, along with *Psilocybe semiglobata* and *Panaeolus papilionaceus*. The annulus is typically located in the lower regions of the stem, or at most mid distance, but not superior. This species is virtually identical to *Psilocybe moellerii,* which has larger spores: 13–14 (16) by 7–8 μ. See also *Psilocybe coprophila* and *Psilocybe subviscida.*

Psilocybe mexicana Heim

COMMON NAMES: nize ("little birds" in the Mazatec language), Mexican liberty caps

Cap: .5–3 cm broad. Conic to campanulate to subumbonate, to convex at maturity, often with a small umbo. Surface viscid to smooth when moist, striate from the margin halfway to the disc. Margin sometimes decorated with fine fibrils. Brownish to deep orangish brown, fading in drying to yellowish, becoming opaque, often with bluish tones from age or where injured. **Gills:** Attachment adnate to adnexed, sometimes sinuate, pale gray to dark purplish brown with spore maturity, typically with whitish edges. **Stem:** 40–125 mm long by 1–3 mm thick, equal to narrowing towards the base, smooth, and hollow. Straw yellow to brownish, darkening with age or where injured. Partial veil thinly fibrillose, whitish, leaving fibrillose veil remnants on the upper regions of the stem. Flesh reddish brown, bruising bluish where injured. **Microscopic features:** Spores dark purplish brown to blackish purple brown in deposit, ellipsoid to subellipsoid in side view, subrhomboid in face view, 8–9.9 (12) by 5.5–7.7 (8) μ. Basidia 4-spored. Pleurocystidia absent or similar to cheilocystidia when near the gill edge. Cheilocystidia 13–28 (34) by 4.4–6.6 (8.8) μ, fusoid-ampullaceous, sublageniform with abbreviated apices, 1.5–2.2 (3.3) μ, occasionally forking. **Habit, habitat, and distribution:** Solitary to gregarious in meadows, often in horse pastures, in soils rich in manure, along field-forest (deciduous) interfaces, most common at 1000–1800 meters in elevation. Found in June through September in subtropical Mexico (Michocan, Morelos, Jalisco, Oaxaca, Puebla, western Xalapa, and

***Psilocybe mexicana*. Moderately to highly active.**

Psilocybe mexicana.

Veracruz), and Guatemala. The trees surrounding meadows hosting this species are sweetgum (*Liquidambar styraciflua*), oaks (*Quercus* sp.), alders (*Alnus* sp.), and others. Also found along river valleys and in pasture lands bordered with sycamore (*Platanus lindeniana*). **Comments:** Moderately potent, Heim and Hofmann (1958) first detected .25% psilocybin and .15% psilocin. Fresh specimens are undoubtedly more potent. This mushroom forms sclerotia in culture that are high in psilocybin and low in psilocin. The sclerotia can be planted in soil and fruited (Stamets and Chilton 1983). From anecdotal reports, the sclerotia of *P. mexicana* hold their potency months after dehydration. I like to think of *P. mexicana* as the liberty cap of Mexico, in that the two species share similar affinities for habitats (manured grasslands), often boast an umbo, are tall, thin, and have deeply translucent-striate margins. See also *Psilocybe tampanensis*.

Psilocybe moellerii Guzman

= *Stropharia merdaria* Fr. sensu Rea

= *Stropharia merdaria* var. *macrospora* (Moller) Singer

Cap: 1–3 (4.5) cm broad. Convex, sometimes slightly umbonate, expanding with age to broadly convex to nearly plane, sometimes slightly depressed in the center. Surface smooth, viscid when wet from a separable gelatinous pellicle, and translucent-striate. Cinnamon brown to dull orangish brown, hygrophanous, fading in drying to yellowish brown to straw colored. **Gills:** Attachment bluntly adnate to subdecurrent, pale brownish to dark purple brown, somewhat spotted from the uneven ripening of spores, with whitish edges. **Stem:** 20–50 (85) mm long by 4–5 (6) mm thick. Equal to slightly enlarged near the base, flexuous, tough, and stuffed with a whitish pith. Surface covered with silky fibrils below the annular zone and smooth above. Partial veil membranous (subfloccose), leaving a fragile membranous annulus that soon deteriorates into a dark, often middle-to-superior annular zone, typically darkened with spores. **Microscopic features:** Spores dark purple brown in deposit, subellipsoid in side view, subhexagonal in face view, 11–14 (16) by 6.6–8.8 μ. Basidia 4-spored. Pleurocystidia absent. Cheilocystidia 18–40 by 4–9 (11) μ, fusoid-ventricose with a flexuous neck, 2.2–4 μ thick. **Habit, habitat, and distribution:** Gregarious, occasionally cespitose, on cow or horse dung, along trails, or in soils enriched with manure. Common in late summer through fall. A temperate species that is widespread throughout the world, *P. moellerii* has been reported from the United States (Oregon and Washington), South America (Peru, Argentina, Venezuela, and Chile), Greenland, Europe (the Czech Republic, Slovakia, France, and Sweden), and New Guinea. Probably more widely distributed throughout temperate regions of the world. **Comments:** Not conclusively shown to be active. Guzman (1983) notes that this comparatively small species prefers colder, boreal climates, while *P. merdaria* is more common in temperate zones. Virtually indistinguishable macroscopically from *P. merdaria, P. moellerii* is one of the more common dung dwellers, often sharing the same dung deposit with *Psilocybe semiglobata* and many *Panaeolus* species.

Psilocybe moellerii. **Not known to be active.**

Psilocybe montana (Fries) Quelet

= *Psilocybe atrorufa* (Schaeffer ex Fries) Quelet

Cap: .5–2 cm broad. Hemispheric at first, rapidly expanding to convex to broadly convex to plane in age, and may be slightly umbonate. Surface smooth, viscid when wet from a separable gelatinous pellicle. Margin translucent-striate from moist, becoming opaque in drying. Dark reddish brown to ochraceous, hygrophanous, fading to light yellowish brown to grayish brown in drying. **Gills:** Attachment adnate to subdecurrent, subdistant, thin to moderately broad. Color light gray brown to very dark reddish brown to violet brown with spore maturity. **Stem:** 15–40 (50) mm long by 1–2 mm thick. Mostly equal to slightly enlarged at the base, often flexuous. Reddish brown overall or nearly concolorous with the cap. Surface dry, smooth or having a few scattered fibrils, soon disappearing and smooth. Partial veil thinly cortinate, soon obscured. **Microscopic features:** Spores dark grayish brown in deposit, subellipsoid in side view and subrhomboid in face view, 5.5–8 (10) by 4–5 μ, thick walled. Basidia 4-spored. Pleurocystidia absent. Cheilocystidia ventricose with elongated apices, sometimes forked, 2–3.5 μ thick, or lageniform (15) 22–45 by 4.4–7.5 (10) μ. **Habit, habitat, and distribution:** Scattered to numerous, found in mossy ground, less common on sandy soils, and common at higher elevations. Reported throughout much of the temperate zones of the world: North America (Arizona, Oregon, Washington, Idaho, Michigan, New Hampshire, New York, Tennessee, Chiapas, Coahuila, Oaxaca,

Psilocybe montana. **Not active.**

Puebla, Morelos, and Veracruz), South America (Colombia, Argentina, and Chile), Europe (Austria, Belgium, Czech Republic, Slovakia, Denmark, Finland, France, Germany, Hungary, Norway, Sweden, and Switzerland), Russia, and Japan. Undoubtedly more widespread. Fruiting from early summer through autumn. **Comments:** Not yet known to be active. *Psilocybe montana* was designated as the type species for the genus *Psilocybe* (Smith 1949) because it exhibits some of the most representative features. This mushroom is almost always found in association with mosslike plants, and may be an obligatory saprophyte on dead root masses. Guzman (1983) notes its association with *Polytrichum* bryophytes, but not *Sphagnum*. Høiland (1978) reported that this is the most common *Psilocybe* in Norway, growing at sea level up into high mountainous elevations. See also *Psilocybe crobula* and *Psilocybe inquilina.*

Psilocybe muliericula Singer and Smith

= *Psilocybe wassonii* Heim

= *Psilocybe mexicana* var. *brevispora* Heim

COMMON NAME: Wasson's *Psilocybe*

Cap: .5–4 (5) cm broad. Conic to conic-campanulate, expanding to convex to broadly convex to nearly plane, often undulating and umbonate. Surface smooth, shortly translucent-striate along the margin. Reddish brown to vinaceous brown, hygrophanous, fading in drying to a pale orangish light brown from the center to the margins, which are often tinged bluish green. Flesh whitish, quickly bruising bluish where injured. **Gills:** Attachment adnexed to sinuate, close, pale-pinkish brown, becoming dark chocolate brown with spore maturity, with similarly colored or slightly paler edges. **Stem:** 20–60 (100) mm long by 2–7 mm thick. Equal to enlarging upwards, covered with fine fibrillose veil remnants, especially when young, smooth in age. Pale pinkish white, bruising bluish where injured. Partial veil finely cortinate, leaving fragile patches of fibrils, soon disappearing. Flesh pinkish brown, tough, hollow, quickly bruising bluish when handled. **Microscopic features:** Spores dark violaceous brown in deposit, subellipsoid to ellipsoid-ovoid, 6–9.9 by 3.8–4 μ. Basidia 4-spored. Pleurocystidia absent. Cheilocystidia 16–24 by 3.5–6 μ, sublageniform with an elongated neck, 1.5–3 μ thick, sometimes forked or mucronate. **Habit, habitat, and distribution:** Gregarious to cespitose in muddy and swampy soils on the walls of ravines in fir (*Abies*) and pine (*Pinus*) woodlands. Found

August through September in Mexico (state of Mexico). **Comments:** Potently active, judging by the strength of the bluing reaction. The name *Psilocybe wassonii* was preempted by Smith and Singer's publication of the name just 24 days prior to the publication of Heim and Wasson's (1958) masterpiece, and given the international rules governing nomenclature, the name first published presides. Since many of Singer's Mexican contacts were provided, in good faith, by Heim and Wasson, this event created a rift between two schools of mycologists that persisted for several decades. Heim and Hofmann (1958) found .02% psilocybin and .01% psilocin in seven-month-old specimens. Undoubtedly, fresh specimens are many orders of magnitude stronger.

Psilocybe natalensis Gartz, Reid, Smith, and Eicker

Cap: 1.4–6 cm broad. Obtusely conic, expanding with age to hemispheric to convex to broadly convex at maturity, occasionally with a small umbo. Yellowish towards the disc, whitish overall, not hygrophanous, and often bluish tinged along the margin, which can be striate. Surface smooth to irregular in places. **Gills:** Attachment bluntly adnate to subdecurrent in age. Buff at first, then dark purplish brown with whitish margins. **Stem:** 40–120 mm long by 2–10 mm thick. Surface smooth, silky white, straight to curved and enlarged near the base, which lacks rhizomorphs. Bruising bluish green where injured, partic-

Psilocybe natalensis. **Moderately to highly active.**

ularly at the base. Partial veil scant to absent, not leaving fibrillose veil remnants. **Microscopic features:** Spores deep violet in deposit, broadly ellipsoid in side view and ovoid in face view, 10–15 by 7–9.4 μ. Basidia 4-spored, although 1-, 2-, and 3-spored basidia observed. Pleurocystidia scattered to abundant, clavate to lanceolate, sometimes mucronate, 18–40 by 10–15 μ. Cheilocystidia lageniform, 16–22 long by 5.5–8 μ broad at the base, narrowing into a neck 3.5–6 μ long and then swelling at the apex. **Habit, habitat, and distribution:** Scattered to gregarious in fertilized soils in pastures. Thus far only reported from Natal, South Africa, at 1500 meters, in January. Undoubtedly more widely distributed. **Comments:** An unusual *Psilocybe,* this new species is nearly white in color in drying—a feature also shared by *Psilocybe aztecorum* and *Psilocybe baeocystis.* Judging by the bluing reaction, this species is probably moderately potent at best. Its aspect, the relatively small cap versus the relatively long stem, delineates this species from many others. Gartz estimates psilocybin and psilocin content to be similar to that of the averages reported by him (1989) for *Psilocybe cubensis.*

Psilocybe pelliculosa (Smith) Singer and Smith

Cap: .5–2 (3) cm broad. Obtusely conic, becoming conic-campanulate with age. Margin translucent-striate and generally not incurved in young specimens. Chestnut brown when moist, then dark dingy yellow to pale yellow in drying (hygrophanous), often with a pallid band along the margin, and frequently tinged olive green in patches. Surface smooth, viscid when moist from a separable gelatinous pellicle. Flesh thin, pliant, and more or less concolorous with the cap. **Gills**: Attachment adnate to adnexed, finally seceding, close, narrow to moderately broad. Color dull cinnamon brown, darkening with spores in age. **Stem:** 60–80 mm long by 1–2.5 mm thick. Equal above, and slightly enlarged at the base. Surface is covered with appressed grayish fibrils, and powdered at the apex. Whitish to pallid to grayish, more brownish toward the base, blue green where bruised or with age. Flesh stuffed with a tough pith. Partial veil thin to obscure or absent. **Microscopic features:** Spores purplish brown in deposit, subellipsoid to subovoid, 9–13 by 5–7 μ. Basidia 4-spored. Pleurocystidia absent. Cheilocystidia 17–36 by 4–7.5 μ, fusiform to lance-shaped, with an elongated neck 1.5–2 μ thick. **Habit, habitat, and distribution:** Scattered

Psilocybe pelliculosa. **Weakly to moderately active.**

to gregarious to cespitose on well-decayed conifer substratum, in mulch, or in soil rich in lignin. Often seen along paths in conifer forests and along abandoned logging roads that are actively being recaptured by alders and firs. Found in mid-to-late fall to early winter throughout the Pacific Northwest and Northern California. **Comments:** Active, although comparatively weak, containing up to .41% psilocybin, no psilocin, and .04% baeocystin, as reported by Beug and Bigwood (1982b) and Repke et al. (1977). *Psilocybe pelliculosa* is nearly identical with *Psilocybe silvatica,* and is distinguished from it by the length of the spores. The conic-shaped cap, the gregarious nature of their fruitings, the fibrillose patches on the stem, and the bluing reaction at the base of the stem are some of the most distinctive features of this species. See also *P. silvatica* and *P. washingtonensis.* This mushroom has a general resemblance, especially at a distance, to *Hypholoma dispersum* (= *Naematoloma dispersum.*) When my kids were toddlers, I would take them mushroom hunting in the Olympic Peninsula. On one November trip, after parking the car, we strolled up an abandoned logging road. Coming down the road were two pot (edible mushroom) hunters. When I asked if there were any good mushrooms up there, they replied, "There's nothing of interest." Less than fifty yards later, we found several thousand *P. pelliculosa.* The patch looked like a well-organized army of mushrooms, standing tall and proud. The pot hunters had walked right through the multitudes, unaware of the potential for a life-changing experience.

Psilocybe physaloides (Bull. ex. Merat) Quelet

= *Psilocybe caespitosa* Murrill

Cap: .5–2.5 cm. Convex to slightly subumbonate at first, soon expanding to broadly convex to nearly plane, often with an undulating margin and subumbonate. Surface smooth, viscid when moist and translucent-striate along the margin, which can be decorated with fibrillose remnants of the partial veil. Chestnut brown to reddish brown, lighter along the margin, hygrophanous, fading in drying to straw yellow or dingy yellow. **Gills:** Attachment adnate to adnexed, or sinuate. Dingy vinaceous brown, becoming dark purplish brown with spore maturity. **Stem:** 15–50 mm long by 1–1.5 mm thick, typically enlarging towards the apex, then narrowing, and enlarging at the base. Rigid, hollow, and covered with patches of fine grayish fibrils, darkening upwards. Partial veil cortinate, deteriorating with age and leaving remnants. **Microscopic features:** Spores violet blackish brown in deposit, subrhomboid in face view, subellipsoid in side view, 6–8 by 4–5.5 μ. Pale yellowish brown to tan with spore maturity. Basidia 4-spored. Pleurocystidia absent. Cheilocystidia 16–28 (33) by 4–7 μ, fusoid to sublageniform with short necks 3.3–4.4 μ thick. **Habit, habitat, and distribution:** Solitary to gregarious, sometimes subcespitose in disturbed soils, soils rich with woody debris, often at field-forest (conifer) interfaces. Common in Europe, also reported from Canada, Greenland, and the northern United States, including Alaska. This mushroom has been collected as far south as

Psilocybe physaloides. **Activity unknown.**

Santa Barbara. Often fruits in the summer through early fall. **Comments:** Not known to be active; not yet analyzed. The silvery sheath of fibrils from a darkening base is a feature that stands out for this mushroom, and is reminiscent of *Psilocybe crobula, Psilocybe inquilina,* and to a lesser degree *Psilocybe montana.*

Psilocybe pseudocyanea (Desmazieres: Fries) Noordeloos

= *Stropharia pseudocyanea* (Desmazieres) Morgan
= *Stropharia albocyanea* (Desmazieres) Quelet

Cap: 1–3 cm, conic-convex, expanding to acutely to bluntly umbonate, whitish with azure to bluish green tinges, fading to straw or cream colored. Surface viscid from a separable gelatinous pellicle. Margin adorned with remnants of the partial veil. Flesh bluish green, fading to azure, then pale bluish green and eventually straw colored. **Gills:** Attachment adnate, pale fawn to purplish, with toothlike edges. **Stem:** 35–70 mm long by 2–5 mm thick, equal, slim, flexuous, bluish green to azure blue to straw colored, soft, easily breaking, pruinose-flocculose particularly towards the apex. Partial veil membranous, leaving a membranous annulus, degrading into an annular zone. **Microscopic features:** Spores purplish brown in deposit, ellipsoid, 7–9 by 4–5 μ. Basidia 4-spored. Cheilocystidia capitate-clavate to lageniform capitate, 24–44 by 4–8 μ by 4–5 at apex when narrowing, or 6–12 μ when swollen. **Habit, habitat, and distribution:** Prefers tall grass in wetlands, marshes, and meadows, or tall shrubs along trails at field-forest interfaces. Known from the British Isles, much of northern Europe, and the Pacific Northwest. Probably more widely distributed. **Comments:** Activity unknown. Due caution is advised. *P. pseudocyanea* is closely related to *Psilocybe aeruginosa* and *Psilocybe caerulea.* All of these species were formerly placed in the genus *Stropharia* but recently transferred to *Psilocybe* by Noordeloos (1995). The smaller spores, soft stem, and wetland habitats are clas-

***Psilocybe pseudocyanea*. Activity unknown.**

sic features. The edibility of this species is not well documented. The bluish color and its close taxonomic relationship to active species makes *P. pseudocyanea* a likely candidate for activity. However, caution is always recommended with these less-studied species that lack an experiential history.

Psilocybe quebecensis Ola'h and Heim

Cap: 1–3.5 cm broad. Nearly hemispheric at first, soon expanding to convex, then becoming broadly convex to plane, and without an umbo. Margin incurved at first and usually not markedly undulated; translucent-striate when moist. Pale straw yellow and often with brownish or tawny hues, becoming more grayish in drying. Bruising bluish when touched or disturbed. Surface smooth, becoming finely wrinkled with age, and viscid when moist. Flesh whitish. **Gills:** Attachment adnate, thin, moderately broad to swollen in the middle. Becoming very dark chestnut brown at maturity, and usually somewhat mottled, with the edges remaining whitish. **Stem:** 20–35 (45) mm long by 1–2.5 mm thick. Brittle, tough, and fibrous. Slightly enlarged at the apex and flared at the base, which is often furnished with rhizomorphs. Yellowish tawny, drying to a distinct grayish yellow, becoming bluish where bruised. Partial veil cortinate, fugacious, and soon disappearing. **Microscopic features:** Spores purplish brown to black in deposit, ellipsoid to subovoid in side and face view, some spores mango shaped, 8–14 (16) by 6–8.8 μ. Basidia 4-spored. Pleurocystidia present, 12–35 by 9–15 μ, very distinctive by their swollen apices, as in *Psilocybe cubensis* and *Psilocybe cyanescens*. Cheilocystidia 18–36 by 5.5–10 μ, fusoid-ampullaceous with an extended neck, 2–3.3 μ thick. **Habit, habitat, and distribution:** Grows in sandy soils, particularly in outwashes of streams, and in the decayed-wood substratum of alder, birch, fir, and spruce in the late summer and fall. Reported from Quebec, specifically in the Jacques Cartier river valley. **Comments:** Moderately active according to Ola'h and Heim (1967).

***Psilocybe quebecensis*. Moderately active.**

P. quebecensis is a classic flood-plain species and is probably more widely distributed than presently reported. This species is not well known to Quebec residents, and I suspect that naturalized colonies could easily be established by those wishing to have *P. quebecensis* growing in their backyard. See also *Psilocybe caerulipes* and *Psilocybe baeocystis.*

Psilocybe samuiensis Guzman, Allen and Merlin

Cap: .7–1.5 cm broad. Convex to conic-convex to campanulate, often umbonate with a small papilla. Surface translucent-striate near the gill edge and viscid when moist from a separable gelatinous pellicle. Chestnut to reddish brown to straw yellow when young, strongly hygrophanous, fading in drying to straw or brownish. **Gills:** Attachment adnate, clay colored, then violaceous brown to chocolate violet with whitish margins. **Stem:** 40–65 mm long by 1–2 mm thick. Whitish to yellowish and covered with fibrillose sheath of veil remnants. Equal to slightly enlarged towards the base. Concolorous with cap, bruising bluish where injured. Partial veil cortinate, leaving a superior, fibrillose annular zone that soon disappears. **Microscopic features:** Spores purplish brown in deposit, rhomboid to subrhomboid, 10–13 by 6.5–8 μ. Basidia 4-spored. Pleurocystidia scattered, ventricose towards the base and sublageniform towards the apex, 16–20 by 5–6.4 μ. Cheilocystidia ventricose at base, lageniform, narrowing to a thinner neck, often forked, 18.5–28 (30) by 5–7 (8) μ. **Habit, habitat, and distribution:** Grows in well-manured, claylike soils in pastures, meadows, or amongst rice paddies. First reported by John Allen from the island Koh Samui off Thailand, this mushroom was first found in early August. The extent of its full fruiting season is not known. *P. samuiensis* may be widely distributed throughout that region. **Comments:** Active. *P. samuiensis* is strikingly similar to the liberty cap (*Psilocybe semilanceata*) and prefers a common habitat with *Psilocybe cubensis* and *Panaeolus cyanescens.* Psilocybin,

Psilocybe samuiensis. **Moderately to highly active.**

psilocin, and baeocystin levels are .73%, .52%, and .05%, respectively (Gartz et al. 1994). Stijve and de Meijer (1993) reported the analysis of a species called *Psilocybe thailandensis* by Guzman and Allen from Koh Samui, Thailand: up to .075% psilocybin and .60% psilocin. (I am unclear as to the relationship between these two taxa.) For more information, consult Guzman, Bandala, and Allen (1993).

Psilocybe semiglobata (Batsch: Fries) Noordeloos

= *Stropharia semiglobata* (Batsch ex Fries) Quelet

Cap: 1–4 cm broad. Obtuse to hemispheric to convex and finally broadly convex in age. Occasionally with an umbo. Light yellow to deep straw yellow when fresh, not hygrophanous. Surface smooth, viscid to extremely viscid to glutinous from a separable gelatinous pellicle. Flesh firm, watery buff, not bruising. **Gills:** Attachment adnate, broad, close to subdistant, with one to two tiers of intermediate gills. Grayish when young, becoming purplish brown with spore maturity, edges remaining pale to whitish. **Stem:** 30–120 mm long by 2–5 mm thick. Equal to slightly enlarged downwards. Surface below annular zone, whitish to nearly concolorous with cap, viscid to extremely viscid to glutinous in the lower two thirds of the stem. Partial veil glutinous, leaving an annular zone of fibrils, soon darkened. Flesh stuffed with a yellowish pith at first, and becoming hollow with age. No bruising reaction. **Microscopic features:** Spores dark purplish brown in deposit, ellipsoid, 15–19 by 8–10 μ. Basidia 4-spored. Cheilocystidia present, narrowly fusoid-ventricose

Psilocybe semiglobata **is difficult to grasp when wet because of its glutinous coating. Not active.**

with a long, flexuous neck, 26–38 by 6–9 µ. Pleurocystidia difficult to find, 32–46 by 9–14 µ. **Habit, habitat, and distribution:** Single to gregarious on cow or horse dung in the spring, summer, and fall. Widely distributed throughout North America and much of the temperate regions of the world. **Comments:** Not active. *Psilocybe semiglobata*, formerly known as *Stropharia semiglobata*, is included here because this mushroom is prolific in habitats also frequented by the active Psilocybes and Panaeoli. I have often found *Psilocybe merdaria, Psilocybe coprophila*, and *Panaeolus papilionaceus* sharing the same habitat with *P. semiglobata*. The natural affinities between these Psilocybes is obvious to all who have collected them. The uniquely glutinous veil distinguishes this species from close relatives.

Psilocybe semilanceata (Fries) Kummer

COMMON NAME: liberty cap, witch's hat

Cap: .5–2.5 cm broad. Conic to obtusely conic to conic-campanulate to campanulate with an acute umbo. Margin translucent-striate, incurved and sometimes undulated in young fruiting bodies, often darkened by spores. Color variable, extremely hygrophanous. Usually dark chestnut brown when moist, soon drying to a light tan or yellow, occasionally with an olive tint. Surface viscid when moist from a separable gelatinous pellicle. **Gills:** Attachment mostly adnexed, close to crowded, narrow. Color pallid at first, rapidly becoming brownish and finally purplish brown with the edges remaining pallid. **Stem:** 40–100 mm long by .75–2 (3) mm thick. Slender, equal, flexuous, and pliant. Pallid to more brownish towards the base, where the attached mycelium may become bluish tinged, especially during drying. Surface smooth overall. Context stuffed with a fibrous pith. Partial veil thinly cortinate, rapidly deteriorating, leaving an obscure evanescent annular zone of fibrils, usually darkened by spores. Often, this zone is entirely absent. **Microscopic features:** Spores dark purplish brown in deposit, ellipsoid, 12–14 by 7–8 µ. Basidia 4-spored. Pleurocystidia few to absent. Cheilocystidia 18–35 by 4.5–8 µ, lageniform with an extended and flexuous neck, often forked. **Habit, habitat, and distribution:** Scattered to gregarious in the fall in pastures, fields, lawns, or other grassy areas, especially rich grasslands grazed by sheep and cows. This mushroom is especially abundant in or about clumps of sedge grass in the damper parts of fields. It can be found west of the Cascades from Northern

Classic forms of *Psilocybe semilanceata*. Highly to extremely potent.

California to British Columbia in the fall to early winter, and to a much lesser degree in the spring along the coastal areas of Oregon and Washington. Reports by Peck of this mushroom from upstate New York are confused (Guzman 1983, 363–364). Redhead (1989) notes several sitings from New Foundland and Nova Scotia. Appears from the late summer through the late fall in the northern latitudes. Reported in grassland habitats in Europe (France, Holland, Italy, Norway, and Switzerland), South Africa, Chile, northern India, Australia, and Tasmania. Guzman, Bandala, and King (1993) reported this species from New Zealand, with specimens collected in the month of May. Johnston and Buchanan (1996) reported *P. semilanceata* only from high-altitude grasslands from the South Island of New Zealand. **Comments:** Moderately active to extremely potent. In the Pacific Northwest, this species is one of the most common of active Psilocybes, and perhaps the easiest for amateurs to identify. Gartz (1993) reported an average of 1% psilocybin, with a range of .2 – 2.37%, the highest psilocybin content yet reported. Cultivated fruitbodies yielded maxima of 1.12% psilocybin, no psilocin, and .21% baeocystin. In separate studies, Gartz (1994) reported up to .98% psilocybin, while Stijve and Kuyper (1985) found a very high concentration—in one specimen—of 1.7%. The high psilocybin and low psilocin probably accounts for its long storage life. The same study showed that this species is also relatively high in baeocystin (.36%). Christiansen et al. (1981) found a variation of psilocybin in dried specimens, .17–1.96%, with the youngest specimens being most potent on a dry-weight basis. Gartz (1986a) also reported that psilocybin content was not adversely affected by the drying process. Since *P. semilanceata* is low in psilocin, a co-indicator, it rarely bruises bluish. In this species, the strength of the bluing reaction is *not* an indication of activity, unlike the majority of potent species, which are comparatively higher in psilocin.

Several distinct varieties of *P. semilanceata* can be encountered. The archetypal variety is easy to recognize macroscopically. However, its many forms can be confusing. Most forms have conic to campanulate caps with a sharp umbo. One unusually large form, probably *P. semilanceata*, with a pronounced incurved margin and contorted bell-shaped cap, occasionally surfaces late in the season in the Pacific Northwest. Perhaps the most unusual form I have found is one that looks dangerously close to a *Galerina* but is in fact a denuded, nearly sporeless, orange form of *P. semilanceata*. With sparse spore production, the color of the overlying translucent flesh shows a yellowish orange color instead of its normal purplish brown. Readers should note that woodland Galerinas can coexist in the same habitat as grassland Psilocybes, especially in lands recently converted to pastures. *Psilocybe strictipes* is a similar species, but lacks the distinctive umbo typical of *P. semilanceata. P. semilanceata* is to temperate grasslands what *Psilocybe mexicana* is to subtropical grasslands. An interesting ecological study by Keay and Brown (1990) illustrates the close relationship of *P. semilanceata* to the rhizomes of grasses. See also *Psilocybe strictipes, Psilocybe pelliculosa, Psilocybe silvatica, Psilocybe subfimetaria,* and *Psilocybe samuiensis.*

Psilocybe serbica Moser and Horak

Cap: 1–3.5 cm broad. Convex to campanulate, often twisted, expanding with age to broadly convex to plane. Surface viscid when moist from a thin gelatinous pellicle, often not separable. Margin translucent-striate when moist, becoming opaque in drying, soon uplifting and irregular in age, bluish tinged or bruising bluish where injured. Reddish brown to brownish yellow, hygrophanous, fading to grayish yellow, ochraceous or straw colored in drying. **Gills:** Attachment adnate to adnexed, purplish brown to chocolate brown with whitish margins. **Stem:** 20–60 mm long by 1.5–5 mm thick. Equal to slightly enlarged towards the base, whitish to whitish with splotchy brown patches, bruising bluish where injured, and covered with white fibrillose patches

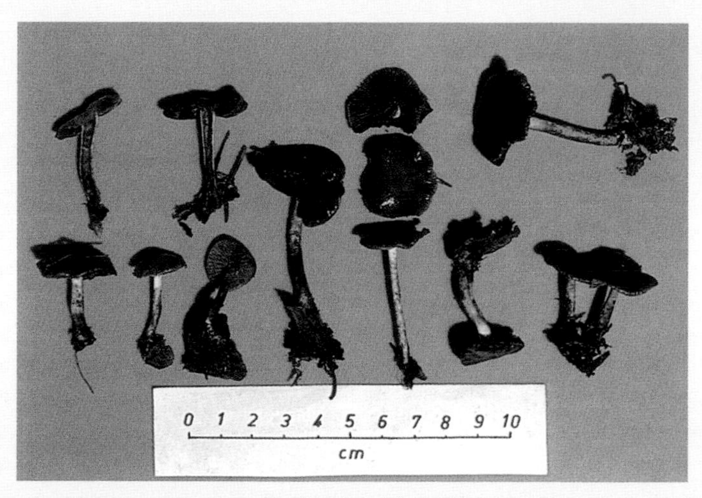

Psilocybe serbica. **Moderately to highly potent.**

below. Partial veil finely cortinate, soon disappearing. **Microscopic features:** Spores dark purplish brown in deposit, elongate-ellipsoid in face and side view, 10–13 by 5.5–7 μ. Basidia 4-spored. Pleurocystidia absent or rare, and usually near the gill edge, mucronate, 16–25 by 4.4–7.7 μ. Cheilocystidia 20–36 by 6–11 μ, lageniform with extended necks 1.5–2.5 μ, arising from cells parallel to gill edge. **Habit, habitat, and distribution:** Reported only from Yugoslavia (Serbia), Slovakia, and the Czech Republic. Growing on rotting wood or in soils rich in woody debris, in deciduous or mixed forests, beneath European beech (*Fagus silvatica*) and/or mixed with firs (*Abies* sp.) **Comments:** Moderately active to highly active. *P. serbica* is taxonomically close to *Psilocybe atrobrunnea* except the latter species prefers mossy areas and does not bruise bluish. See also *Psilocybe liniformans* and *Psilocybe cyanescens.*

Psilocybe silvatica (Peck) Singer and Smith

Cap: .8–2.5 cm broad. Obtusely conic to campanulate, and often with an acute umbo. Tawny dark brown when moist, fading to pale yellowish brown or grayish brown. Surface smooth, viscid when moist from a thin gelatinous pellicle that is barely separable, if at all. **Gills:** Attachment adnate to adnexed, close to subdistant, narrow to moderately broad. Color dull grayish brown to cinnamon brown to smoky brown at maturity, with the edges remaining whitish. **Stem:** 20–80 mm long by 1–3 mm thick. Equal to slightly enlarged at the base, brittle, tubular, and somewhat flexuous. Pallid to brownish beneath a whitish fibrillose covering. Partial veil

Psilocybe silvatica. **Weakly to moderately active.**

poorly developed, cortinate, thin to obscure, and soon absent. **Microscopic features:** Spores dark purplish brown in deposit, 6–9.5 by 4–5–5.5 μ from 4-spored basidia; sometimes 2-spored. Pleurocystidia absent. Cheilocystidia 24–40 by 4.4–7 (8.8) μ, fusoid ventricose to lageniform, with a long flexuous neck, 1.6–2.2 μ thick. **Habit, habitat, and distribution:** Gregarious but not cespitose on wood debris or on wood chips or in well-decayed conifer substratum in the fall. Known from Ontario, the Pacific Northwest, Michigan, New York, and northern Europe. **Comments:** Contains psilocybin and/or psilocin. Estimated not to be potent. Differing from *Psilocybe pelliculosa* in the length of the spores and cheilocystidia. The colonies of *P. silvatica* I have found have had consistently more yellow in the caps compared to *Psilocybe pelliculosa*, its closest relative.

Psilocybe squamosa (Persoon ex Fries) Orton

Cap: 3–8 cm broad. Conic or obtusely conic when young, with an incurved margin soon becoming broadly campanulate to convex, and often with a conic umbo. Viscid when moist from a separable gelatinous pellicle, but soon drying. Yellowish brown to orangish brown overall. At first adorned with small squamules (scales) along the margin, but soon smooth and free of veil remnants. Flesh relatively thin and watery brown when wet. **Gills:** Attachment adnate to uncinate, close to subdistant, moderately broad, with two to three tiers of intermediate gills inserted. Pallid bluish gray, then dark grayish brown to purple brown when fully mature. **Stem:** 60–120 mm long by 4–8 mm thick. Equal to slightly enlarged and curved at the base, hollow, and somewhat fragile. Whitish towards the apex and sordid brown to dense buff below the annulus. Surface covered with evanescent pallid to brownish recurved scales, often with orange-buff rhizomorphs protruding about the base. Partial veil membranous, fragile, leaving a superior membranous annulus, striated on the top side, often hanging broken around the stem or entirely absent in age. **Microscopic features:** Spores

***Psilocybe squamosa*. Not active.**

dark grayish violet brown in deposit, subellipsoid, 11–15 by 7–8 μ with a central germ pore. Basidia 4-spored. Cheilocystidia 36–66 by 3.3–6 μ, filamentous to sublageniform, with an apex adorned with a glutinous mass. Pleurocystidia absent. **Habit, habitat, and distribution:** Solitary to scattered in the late summer and fall in meadows and mixed conifer and alder woods. Known from the Pacific Northwest, Minnesota, and California. It is likely this species is widely distributed across the continent. **Comments:** There are conflicting reports on the edibility of *P. squamosa*; caution is definitely recommended. Once a *Stropharia,* this species lacks the chrysocystidia once typical of that genus. Both the squamules of the cap and the annulus are temporal features, and can soon disappear after heavy rains. See also *Psilocybe thrausta,* which is very similar but differs in the reddish coloration of the caps.

Psilocybe strictipes Singer and Smith

= *Psilocybe callosa* (Fries ex Fries) Quelet sensu auct., sensu Guzman (1983)
= *Psilocybe semilanceata* var. *obtusa* Bon.
= *Psilocybe semilanceata* var. *microspora* Singer

Cap: .5–3 cm broad, conic at first, expanding to convex, campanulate, and eventually broadly convex, and typically not sharply umbonate but may have a low umbo. Surface smooth, translucent-striate near the margin, which may have slight remnants of the veil and viscid when moist from a separable gelatinous pellicle. Dark grayish brown to cinnamon brown, fading to straw or light yellow in drying. Flesh sometimes bruising bluish when injured. **Gills:** Attachment adnate, sometimes subdecurrent, and tearing free from the stem in drying. Chocolate brown with whitish edges when mature. **Stem:** 40–70 (130) mm long by 2–3 mm thick. White to yellow to yellowish brown. Equal, straight to flexuous, typically tough, cartilaginous, and decorated with fibrillose patches, veil remnants, and basal mycelium that can bruise bluish. (Base is not adorned with cordlike rhizomorphs.) Partial veil thinly cortinate, fragile, and rarely leaving an annular zone on the upper regions of the stem. **Microscopic features:** Spores dark purple brown in deposit, subellipsoid to suboblong, 10–12 by 5.5–8 μ. Basidia 4-spored. Pleurocystidia absent. Cheilocystidia 21–45 by 7–10 μ, lageniform with an extended neck, 2–3.5 μ thick. **Habit, habitat, and distribution:** Fruits in the

Psilocybe strictipes. **Moderately to highly active.**

late summer to late fall in the Pacific Northwest, England, northern and central Europe (Czech Republic, France, Germany, Holland, Slovakia, Sweden), Siberia, and Chile. Typically found in rich, grassy areas such as lawns, along roadsides, and in fields—but not on dung, although common in fields with and without manure. **Comments:** Chemical analyses not available. Estimated to be moderately active to potent, judging from personal bioassays, and probably low in psilocin, because of the limited bluing reaction. *P. strictipes* is a slender, grassland species, thought to be an intermediate form, bridging *Psilocybe semilanceata* and *Psilocybe pelliculosa*, two taxa that are very similar in general appearance except for habitat preferences and/or microscopic details. *P. strictipes* has had a very confused history. Guzman (1995), following Redhead (1985) and Watling and Gregory (1987), attempted to clear up the long-standing confusion surrounding this species and its well-known synonym, *Psilocybe callosa*. (The original *Agaricus callosus* Fr. is not related to the mushroom described here and is actually synonymous with *Panaeolus papilionaceus* Bull. ex: Fr. Quelet.)

The modern concept of *Psilocybe callosa* became subordinate to a new taxon, which Singer and Smith (1958a) originally proposed as *P. strictipes*. Guzman (1983) had made *P. strictipe's* subordinate to *Psilocybe callosa* in his monograph. Upon reevaluation, Guzman (1995) reaffirmed *P. strictipes* as the proper name. Mixed collections resulted in this species being further confused with *Psilocybe baeocystis*, to which it bears little resemblance. Furthermore, the preferred habitat for *P. strictipes* is grasslands or rich soils, not the woodlands that Singer and Smith (1958a) had described. Their line drawings of *P. strictipes* show two distinct forms: one mycenoid resembling the closely related *Psilocybe semilanceata,* and an isolated drawing, more collyboid in shape, showing a mushroom atypical to the first form. Guzman (1983) writes that the specimen used by Smith to make the line drawing was actually *Psilocybe baeocystis*. Additionally, in their original description they indicate that the collection number assigned to the original type was erroneous, which perhaps related to the original confusion and the two decades of confusion that followed. Virtually all the field guides published since 1958, including mine (1978), erroneously describe *P. strictipes*.

For most field hunters, the grassland habitat narrows the field of candidate possibilities. The absence of a sharp umbo and its thinly fleshed cap are two macroscopic features that delineate this species from its closest ally, *Psilocybe semilanceata,* with which it is often confused. Guzman (1983, 17) noted that "the form of the pileus is of taxonomic value in *Psilocybe*. *P. semilanceata* is distinguished from *P. callosa* (= *P. strictipes*) on the papilla in the first and in the convex to the more or less subumbonate pileus in the latter."

With many grassland species, the length of the stem is usually a direct response to the height of the grass through which its arises. The stem base is typically tightly attached to dead, thatched grass. Microscopically, *P. strictipes* has smaller and narrower spores than *Psilocybe semilanceata*. The name *strictipes* refers to the tough or hardened texture of the stem, especially the base, in drying.

A variety of *P. strictipes* grows abundantly in western Oregon in close association with highland bentgrass (*Agrostis tenuis*) where thousands of acres are dedicated to the commercial cultivation of grass seed—a major industry in that region. The prolific fruitings of *P. strictipes* in these grasslands and the subsequent distribution of spore-dusted seeds represents a huge launching platform of germ plasm to faraway lawns, golf courses, and institutions of higher learning. The potential distribution of this species through the commercial distribution of lawn seed is mind-boggling. *P. strictipes* is likely to be much more common than presently realized.

Psilocybe stuntzii Guzman and Ott

= *Psilocybe pugetensis* Harris

COMMON NAMES: Stuntz's *Psilocybe*, Stuntz's blue legs, blue ringers

Cap: 1.5–5 cm broad. Cap obtusely conic at first, soon expanding to convex to broadly convex-umbonate to nearly flattened or plane, with the margin uplifting in very mature fruiting bodies. Margin translucent-striate halfway to the disc when moist; decurved, then straightening, and finally elevated, undulating, and often eroded in extreme age. Dark chestnut brown, lighter towards the margin, which is often olive greenish; hygrophanous, fading to a more yellowish brown to pale yellow in drying. Some varieties tend to be more olive yellowish brown and are not very hygrophanous. Context relatively thin, watery brown or nearly concolorous with the cap. Surface viscid when moist from a separable gelatinous pellicle. **Gills:** Attachment adnate to adnexed, close to subdistant, moderately broad, with three tiers of intermediate gills. Color pallid in young fruiting bodies, soon becoming more brownish and eventually very dark brown with spore maturity. **Stem:** 30–60 mm long by 2–4 mm thick. Subequal, slightly enlarged at the apex and often curved, twisted and inflated at the base. Dingy yellow to pale yellowish brown. Surface dry, covered with pallid appressed fibrils to the annulus, and smooth above. Context stuffed with a fibrous whitish pith. Partial veil thinly membranous, typically streaked bluish, leaving a fragile membranous annulus as the cap expands, which soon deteriorates into

Psilocybe stuntzii **var. *stuntzii*. Weakly to moderately active.**

a fairly persistent annular zone darkened by spores. Stem often with rhizomorphs protruding about the base. **Microscopic features:** Spores dark purplish grayish brown in deposit, subellipsoid in side view, subrhomboid in face view, (8) 9–10.5 (13.5) by 5.5–7.5 µ. Basidia 4-spored. Pleurocystidia absent. Cheilocystidia 22–30 by 4.4–6.6 µ, lageniform, fusoid-ampullaceous, or fusiform-lanceolate with an elongated and flexuous neck 1–2.2 µ thick. **Habit, habitat, and distribution:** Grows in gregarious to subcespitose clusters on conifer wood chips, in soils rich with woody debris, in newly placed lawns and fields, along roads, paths, in gardens. Common in the fall to early winter, and to a minor degree in the spring. Abundant within ninety kilometers of coastal regions, especially in Oregon, Washington, and British Columbia. Often, this mushroom fruits in prodigious colonies. **Comments:** Weakly to moderately active. Beug and Bigwood (1982b) reported a range of 0–.36% psilocybin, 0–.12% psilocin. Repke et al. (1977) reported .02% baeocystin. By weight, *P. stuntzii* is one of the less potent of the bluing Psilocybes. The most characteristic feature of this species is its whitish, partial veil that bruises bluish or is bluish tinged. This species often grows in colonies of great numbers and was named in honor of Dr. Daniel Stuntz, who made the type collections. The field variety of this species, slender and paler, is *P. stuntzii* var. *tenuis*. The *P. stuntzii* group encompasses a great variety of forms growing in varied habitats. Two taxonomically similar species are *Psilocybe caeruleoannulata*, reported from Uruguay and Brazil, and *Psilocybe jacobsii,* collected in Oaxaca. *Galerina autumnalis,* a deadly mushroom, has the same overall appearance to *P. stuntzii* and would look identical if you were color blind. The orangish brown cap and rusty brown spores are the major differences visible to the unaided eye. See also *Psilocybe subaeruginascens.*

Psilocybe subaeruginascens Hohnel

= *Psilocybe aerugineomaculans* (Hohnel) Singer and Smith

Cap: 1–5.8 cm broad. Conic to convex or campanulate, to broadly subumbonate but not papillate, eventually broadly convex to plane and uplifting in age. Surface viscid when moist, translucent-striate along the margin, soon drying. Orangish brown to olive brown to gray greenish brown, hygrophanous, fading in drying to dull yellow orange to straw colored. Flesh white to concolorous with cap, soon bruising bluish where injured. **Gills:** Attachment broadly adnate to adnexed, sometimes decur-

Psilocybe subaeruginascens. Moderately active.

rent, crowded, and sometimes forking. Grayish brown to yellowish brown, eventually (with spore maturity) dark brown and often slightly mottled. Edges concolorous, bruising bluish where bruised. **Stem:** 30–60 mm long by 1.5–3 mm thick. Equal to slightly enlarged near the base, which is often adorned with radiating white rhizomorphs that bruise bluish when injured. Surface and flesh whitish to concolorous with the cap, soon bruising bluish. Partial veil membranous, well developed, leaving a persistent membranous annulus in the superior regions of the stem, white in color until bruised, and then bluish, and usually dusted purplish brown from spores. Torn and fragile in age. **Microscopic features:** Spores dark purplish brown to dark violet brown in deposit, rhomboid to subrhomboid in face view, subellipsoid in side view, 7.7–12 by 6.6–8.5 µ. Basidia 4-spored, rarely 1-, 2-, or 3-spored. Pleurocystidia fusoid-ventricose with a blunted end, 2.2–3.3 µ thick. Cheilocystidia fusoid-ventricose to sublageniform, 16–33 by 4.4–5.9 µ, with a neck 2.5–4 µ thick. **Habit, habitat, and distribution:** Grows gregariously to cespitose on soils enriched with woody debris, in wood chips, and in wood chips mixed with horse dung. Frequently found along trails or roadsides bordering deciduous forests. Fruiting April to July in temperate southern Japan and subtropical Indonesia. Probably extensively distributed between these two localities. **Comments:** Moderately potent. Koike et al. (1981) detected psilocybin and psilocin. This squat, collybioid, and annulate *Psilocybe* is unique to Asia. Guzman (1983) notes that Singer and Smith's attempted synonymy of *Psilocybe subaeruginascens* with *Stropharia venenata* (= *Psilocybe venenata*) was in error. See also *Psilocybe subfimetaria* and *Psilocybe stuntzii.*

Psilocybe subaeruginosa Cleland

Cap: 1.5–5 cm broad. Conic to convex at first, soon expanding to broadly convex with a small umbo. Surface smooth, translucent-striate along the margin, pale brown to dark brown, hygrophanous, fading in drying to pallid brown to dingy grayish white. **Gills:** Attachment adnate to adnexed, and with age may separate from the stem, leaving parallel, longitudinal gill fragments on the upper reaches. Soon smoky brown to purplish brown to dark purplish chocolate brown, edges concolorous. **Stem:** 50–125 mm long by 2–5 mm thick. Equal to narrowing towards the apex to slightly swollen at the base, from which white mycelium radiates. Hollow, cartilaginous, surface adorned with fine fibrils, finely striate and mealy towards the apex. Surface whitish, silky, with grayish brown streaks, flesh brownish, bruising bluish where injured. Partial veil white, finely cortinate, soon disappearing and leaving fragile traces in the upper regions of the stem. **Microscopic features:** Spores purplish brown in deposit, subellipsoid in both side and face view, 13–15 by 6.6–7.7 μ. Basidia 4-spored, infrequently 2-spored. Pleurocystidia 22–47 by 6–16.5 μ, fusoid-ventricose, mucronate with an elongated neck 2–4.5 μ thick. Cheilocystidia similar in form to pleurocystidia, 17–29 by 5.5–11 μ. **Habit, habitat, and distribution:** Solitary to gregarious in complex habitats such as soils rich in woody debris, decaying piles of leaves and twigs, sandy woody soils, gardens, and amongst bark chips from pine (*Pinus radiata*). Found from May through August. Known only from Australia and Tasmania. **Comments:** Moderately to potently active, judging by the bruising reaction. No analyses are known to me. This species is in the center of a constellation of close relatives, including *Psilocybe australiana, Psilocybe eucalypta,* and

***Psilocybe subaeruginosa*. Moderately to potently active.**

Psilocybe tasmaniana. A study by Chang and Mills (1992) sought to show synonymy between these taxa but, upon close reading of their work, some doubt remains if they had the true *P. subaeruginosa* and were making valid comparisons. *P. subaeruginosa* has pigmented pleurocystidia and is described as "chocolate brown," features Chang and Mills admit to not finding in any of the collections they studied. The chocolate-brown cystidia of *P. subaeruginosa* differentiates this taxon from the mushrooms mentioned above, all of which have hyaline cystidia (Guzman, Bandala, and King 1993). Hence, I think sufficient doubt is cast on their arguments for conspecificity. Better studies are needed.

Psilocybe subcaerulipes Hongo

Cap: 2–3 cm broad. Conic to campanulate expanding in age to convex to subumbonate. Surface smooth viscid when wet and striate near the margin. Ochraceous to yellowish brown, sometimes olive toned, hygrophanous from the disc, fading in drying to light straw colored, bruising dark bluish where injured. **Gills:** Attachment adnate to sinuate, pale brownish to purplish brown to deep violet with concolorous edges. **Stem:** 40–70 mm long by 3–5 mm thick. Equal to slightly enlarging and often curved at the base, flexuous, hollow, and featuring a coating of whitish fibrils in the lower regions. Whitish at first, soon ochraceous or reddish brown, especially towards the base. Partial veil cortinate, fragile, and usually not leaving a fibrillose annular zone on the stem. Flesh bruising bluish where injured. **Microscopic features:** Spores dark

***Psilocybe subcaerulipes*. Probably moderately to highly active.**

purplish brown in deposit, subellipsoid in side view; nearly ovoid in face view, 5.5–7.5 (8.5) by 3.3–4.4 (5.5) μ. Basidia 4-spored. Pleurocystidia absent. Cheilocystidia 15–22 by 5–7.5 μ, fusoid ventricose with an abbreviated or elongated neck 1–1.5 μ thick. **Habit, habitat, and distribution:** Gregarious to cespitose, fruiting from May to September in Japan (Otsu City, Shiga-Prefecture) in soils covered with mosses or grasses in open forests, often under pines, particularly *Pinus densiflora*. **Comments:** Potency unknown, likely to be moderately to highly active. Microscopically nearly identical to *Psilocybe caerulipes* but differing in the size of the spores and its ecological distribution. Furthermore, Guzman (1983) notes that *Psilocybe caerulipes* may be conspecific with *Psilocybe muliercula* (= *Psilocybe wassonii*) as they are difficult to separate taxonomically. Interfertility studies would clear up the question of conspecificity. See also *Psilocybe venenata*.

Psilocybe subfimetaria Guzman and Smith

Cap: .5–2 cm broad. Conic to campanulate, expanding with age to broadly campanulate but not sharply umbonate. Surface smooth, viscid when moist from a separable gelatinous pellicle, and translucent-striate near the margin. Ochraceous brown to olive brown, hygrophanous, fading in drying to a straw color. Flesh whitish, bruising bluish where injured. **Gills:** Attachment adnexed, clay becoming violaceous brown, with whitish edges. **Stem:** 25–45 mm long by 2–3 mm thick, equal to slightly enlarged towards the base, white to dingy white or pallid, bruising bluish where injured. Well-developed partial veil cortinate, usually leaving a fibrillose annular zone that can approach a true annulus in the upper regions of the stem. Flesh whitish, bruising bluish. **Microscopic features:** Spores dark violaceous brown in deposit, subellipsoid in face view, and subellipsoid, sometimes irregularly so, in side view, 10–14 by 6.6–7.7 μ. Basidia 4-spored. Pleurocystidia absent. Cheilocystidia 20–28 by 5–6.6 μ, fusoid-ventricose or more frequently lageniform with an extended neck 1–2 μ broad. **Habit, habitat, and distribution:** Solitary to gregarious on dung, primarily in grassy areas. Reported from locations near Siltcoos Station, Oregon, and Vancouver, British Columbia, in October through November, and near Chovellen, Chile, in August. Probably much more extensive in its distribution than presently realized. I have found this species in habitats also supporting *Psilocybe semilanceata* and *Psilocybe liniformans*. **Comments:** Active, but I do not

***Psilocybe subfimetaria*. Active, but potency unknown.**

know how potent. This mushroom resembles *P. semilanceata* and is definitively separated from it by microscopic features. Macroscopically, this species is distinguished from *P. semilanceata* if the annulus of *P. subfimetaria* persists into maturity. Furthermore, *P. semilanceata* is sharply umbonate, while *P. subfimetaria* is not. *P. subfimetaria* is a relatively rare species and is very close microscopically to *Psilocybe fimetaria*, from which it is separated by spore size. Also, *P. subfimetaria* often grows directly out of dung, while *P. semilanceata* prefers grasses. (Manured grasslands can make this judgment difficult.) The overall aspect, fairly persistent annular zone, and nonpapillate cap can help narrow the field of candidates to *P. subfimetaria*. See also *Psilocybe stuntzii* var. *tenuis*.

Psilocybe subviscida (Peck) Kauffman

Cap: .5–2 cm broad. Campanulate expanding with age to convex-umbonate or broadly convex while retaining an obtuse umbo, often variable in form, and at maturity to nearly plane. Margin translucent-striate when moist and sometimes decorated with flecks of veil remnants. Yellowish to chestnut brown to reddish brown, fading in drying to pale grayish yellow and usually with the umbo remaining reddish brown. Surface viscid to subviscid when moist from a separable gelatinous pellicle, soon drying. **Gills:** Attachment adnate, subdistant, and broad. Whitish at first, soon becoming dark brownish. **Stem:** 20–40 mm long by 1–2 mm thick. Equal to tapering downwards near the base. Surface covered at first with fine whitish fibrils. Partial veil thin to obscure, leaving an

***Psilocybe subviscida*. Not active.**

evanescent annular zone of fibrils usually darkened by spores when present. **Microscopic features:** Spores dark purplish brown in deposit, 6–8.5 by 4–5.4 μ. Basidia 4-spored. Pleurocystidia absent. Cheilocystidia 20–50 by 5–8 μ, ventricose, then lageniform, with a flexuous neck 2–3.5 μ thick. **Habit, habitat, and distribution:** Usually found growing in grassy areas, in well-manured grounds or in dung. Also reported growing in mossy areas and in decayed conifer substrata. Grows in the late spring to summer, reported from the United States (Washington, Oregon, Michigan, New York) and Scotland (Shetland). Thought to be widely distributed. **Comments:** Not active. Since this species can grow on such a wide spectrum of habitats, I would expect that its actual distribution is far more extensive than presently reported. The specimens I have seen have had rubbery textures. See also *Psilocybe coprophila, Psilocybe montana, Psilocybe merdaria, Psilocybe moellerii.*

Psilocybe tampanensis Guzman and Pollock

Cap: 1–2.4 cm broad. Convex, expanding with age to plane or even slightly umbilicate. Surface smooth, subviscid when moist, soon dry, not striate, ochraceous brown to straw brown, hygrophanous, fading in drying to light straw to yellowish gray, with slight bluish tones. **Gills:** Attachment adnexed, soon brownish to dark violet brown with paler edges. **Stem:** 20–60 mm long by 1–2 mm thick. Equal to enlarging near the base, covered with fibrillose patches near the apex and adorned with whitish mycelium at the base, sometimes bluish toned. Yellowish brown to reddish brown overall. Flesh whitish to yellowish, bruising bluish where injured. Partial veil cortinate, soon disappearing, and generally not leaving a fibrillose annular zone on the stem. **Microscopic features:** Spores purplish brown in deposit, subellipsoid in side view, subrhomboid in face view, 8–10 (12) by 6–8.8 μ. Basidia 4-spored. Pleurocystidia absent. Cheilocystidia 16–22 by 4–9 μ, lageniform, with a flexuous, extended neck 2–3 μ thick, irregularly branching infrequently. **Habit, habitat, and distribution:** Reported from Florida and Mississippi in the fall. **Comments:** Moderately potent. Analyses of the sclerotia by Gartz et al. (1994) found up to 0.68% psilocybin and .32% psilocin, respectively. This species, to the best of my knowledge, has only been collected in the wild twice. It was first discovered by Stephen Pollock and Gary Lincoff outside of Tampa, Florida. During a tearfully boring taxonomic conference, and feeling socially unwelcome, Pollock and Lincoff left for a nearby mushroom hunt. Crossing a sand dune, they found a lone specimen of a mushroom that neither of them recognized. Later that day, the specimen yielded purple-brown spores and bruised bluish. Pollock cloned the less-than-spectacular specimen, which soon gave rise to a pure culture—a culture that is in wide circulation today. Since then, there have been no further sightings of wild *P. tampanensis* in Florida. This happenstance finding, and its survival through cultivation, is typical of many of the rare *Psilocybe* discoveries. Stamets and Chilton (1983) first described and illustrated methods for the cultivation of *P. tampanensis* and its sclerotia. See also *Psilocybe mammillata* and *Psilocybe mexicana.*

Psilocybe tampanensis **(cultivated). Moderately to highly active.**

Psilocybe tasmaniana Guzman and Watling

Cap: 1–2 cm broad. Convex to subcampanulate, but not umbonate or papillate. Surface smooth, subviscid when moist, striate only near the margin, which is often adorned with whitish remnants from the veil. Tawny orangish brown, hygrophanous, fading in drying to a dull straw color or dingy yellow. **Gills:** Attachment adnate, becoming purplish brown with spore maturity, with the edges remaining whitish. **Stem:** 40–65 mm long by 1–2 mm thick. Equal overall. Surface silky fibrillose, white to nearly concolorous with the cap. Flesh bruising bluish where injured, especially at the base. Partial veil well developed, finely cortinate, white, soon disappearing. **Microscopic features:** Spores dark purplish brown in deposit, ellipsoid to subellipsoid in side view and subovoid in face view, 10–13 (15) by 7–8.8 μ. Basidia 4-spored. Pleurocystidia 19–24 by 5.5–9.9 μ, fusoid-ventricose with an abbreviated apex, 1.6–2.7 μ broad. Cheilocystidia fusoid ventricose to sublageniform, with extended neck 5–11 by 1.6–3.3 μ. **Habit, habitat, and distribution:** Grows solitary or gregariously in April and May on dung (possibly kangaroo) or in dung-enriched debris in open areas within *Eucalyptus* forests. Reported from Tasmania, New Zealand, and Australia. **Comments:** Active, potency unknown. This species is closely related to *Psilocybe subaeruginosa* and *Psilocybe cyanescens*. Chang and Mills (1992) tried to make *P. tasmaniana, Psilocybe australiana,* and *Psilocybe eucalypta* subordinate synonyms of *Psilocybe subaeruginosa*. For reasons described on pages 91–92, I am skeptical of their interpretations. Furthermore, they made conclusions about the identification of mushrooms that they called *P. tasmaniana* that are not referenced against any type listed by Guzman and Watling, which is surprising considering the strong taxonomic proposals they put forward. If their identifications were in error, and they then "proved" synonymy between the collections, the analyzed collections may indeed be biologically compatible because they incorrectly delimited one species into the other aforementioned taxa. Interestingly, Johnston and Buchanan (1996), while endorsing most of the proposed synonymy of Chang and Mills, selectively exclude *P. tasmaniana* as a species found in New Zealand (nor do they list it as a synonym!) although Chang and Mills (1992) references a collection from Taranak, New Zealand. Clearly, more studies are needed, using better reference standards and systematics, to adequately answer the questions raised here.

Psilocybe thrausta (Schulzer ex Kalchbremer) Orton
= *Psilocybe squamosa* var. *thrausta* (Schulzer ex Kalchbremer) Guzman
= *Stropharia thrausta* (Schulzer ex Kalchbremer) Bon
COMMON NAME: big red scaly *Stropharia*

Cap: 3–7 cm broad. Obtusely conic at first, soon becoming convex to broadly convex, and finally nearly plane with or without an umbo. Viscid when moist from a gelatinous pellicle that is usually separable, soon drying. Surface adorned with whitish, floccose scales, especially near the margin. Orangish red to reddish brown or brick red. Margin initially ornamented with small scales, soon becoming smooth. **Gills:** Attachment adnate to adnexed, or sinuate, sometimes uncinate, close to subdistant, moderately broad, with two to three tiers of intermediate gills. Pallid gray at first, soon becoming grayish brown and eventually dark purplish brown when fully mature. **Stem:** 50–100 (120) mm long by 3–7 (8) mm thick. Nearly equal to swollen and often curved at the base. Hollow in age. Pallid towards the apex and more brownish below. Covered at first with small brown to reddish brown floccose scales to the annulus, and usually with orangish rhizomorphs protruding about the base. Partial veil membranous, fragile, leaving a superior membranous annulus often absent in age. **Microscopic features:** Spores grayish purplish brown in deposit, close to *Psilocybe squamosa* in size, but with an eccentric germ pore, 11–14 by 6.6–8.5 μ. Basidia 4-spored.

Psilocybe thrausta. **Not active.**

Pleurocystidia absent. Cheilocystidia (36) 44–66 by 3.3–6 µ, sublageniform to filamentous, submucronate, with an extended and flexuous neck, 3–4 µ thick. **Habit, habitat, and distribution:** Scattered in the fall in decayed wood substratum or wood debris. Reported from the United States (the Pacific Northwest, New York, Maryland), Japan, and northern to central Europe. Probably more widely distributed across the world. **Comments:** Not active, edible according to some, but not palatable. Høiland (1978) detected no psilocybin. This species was once considered a variety of *Stropharia squamosa* (now a *Psilocybe)* and is very similar to it in appearance, differing in the cap coloration.

Psilocybe venenata (Imai) Imazecki

= *Psilocybe fasciata* Hongo

COMMON NAMES: the false deadly *Psilocybe*, the bamboo *Psilocybe*

Cap: 1–6 cm broad. Initially conic, soon convex to subumbonate to nearly plane in age, often depressed in the center with slightly upturned margins. Chestnut to cinnamon brown to olive brown, then pinkish buff to light blonde brown in drying, often grayish green, rarely whitish, readily bruising bluish where injured. Surface smooth, viscid when moist, translucent-striate near the margin, which is often adorned with minute fibrillose veil remnants, especially when young. **Gills:** Attachment adnate to adnexed, whitish at first, soon light grayish to dingy yellowish, and eventually purplish brown to grayish violet brown with spore maturity. Edges pallid to whitish fringed. **Stem:** 40–90 mm long by 2–6 (9) mm thick, equal to uneven, fibrous, silky white, enlarged near the base from thick white rhizomorphs radiate. Partial veil thickly cortinate to nearly floccose, leaving a fragile fibrillose annulus or annular zone in the superior regions of the stem, soon disappearing or dusted with purplish brown spores. Flesh whitish, bruising azure blue where bruised. **Microscopic features:** Spores dark purplish brown in deposit, subellipsoid, (8) 9.9–12 (14) by 5.5–7 (8.8) µ. Basidia 4-spored. Pleurocystidia absent. Cheilocystidia 17–30 (36) by 4.4–7.5 µ, fusoid-ventricose to lageniform with an extended and flexuous neck, 1–2.5 µ thick. **Habit, habitat, and distribution:** Gregarious to cespitose in the summer and autumn in disturbed habitats, in soils rich with lignicolous debris, and in deciduous or bamboo forests. Also found in composting soils rich with mixtures of rice hulls, straw or manure. Sometimes grows in lawns, along roadsides, or along interfaces in Japanese gardens.

Psilocybe venenata. **Probably moderately to highly potent.**

Reported only from Japan, but I suspect that this species is probably more widely distributed. **Comments:** Probably potent, although no analyses are known to me. *P. venenata* is a misnomer, as no deaths have actually occurred (despite erroneous reports) (Ott 1993). The symptoms produced are typical of other potent psilocybin species. For most collectors, the lighter color, strong bluing reaction, cespitose habit, and locality narrow the field of candidates to this species. Given its preference for habitats, this species is likely to be cultivated outdoors amongst bamboo or in gardenlike settings in a similar fashion to *Psilocybe cyanescens* and *Psilocybe azurescens*. See also *Psilocybe argentipes* and *Psilocybe subcaerulipes*.

Psilocybe washingtonensis Smith

Cap: 1–2 cm broad. Obtusely conic to convex. Margin incurved at first, then straight. Deep walnut brown, lighter towards the margin, hygrophanous, fading to dull cinnamon or pale pecan in drying. Surface smooth, though the margin may be ornamented with faint remnants of the veil, viscid when moist from a gelatinous pellicle, separable only in shreds when wet. Context thin, pliant, and nearly concolorous with the cap when moist. **Gills:** Attachment adnate or subdecurrent, close to subdistant, with two or three tiers of intermediate gills. Color slightly darker than the cap at maturity. **Stem:** 30–50 mm long by 1.5–2.5 mm thick. Equal, tubular, and pliant. Concolorous with the cap, but becoming blackish brown in age from the base upwards. Surface covered with grayish fibrils and adorned with grayish mat of mycelium around the base. Flesh dark brown, fading in drying to pallid tan. Partial veil thinly cortinate, leaving fibrillose patches on the stem and remnants along the cap margin. **Microscopic features:** Spores purplish brown in deposit, ellipsoid to slightly ovoid, 6–7.5 (8) by 4–4.5 μ. Basidia 4-spored. Pleurocystidia 38–56 (64) by 9–12 μ, narrowly fusoid-ventricose to lageniform with expended necks. Cheilocystidia 18–38 by 7–12 μ, variable, fusoid-ventricose, often mucronate, or clavate to capitate with an abbreviated neck. **Habit, habitat, and distribution:** Scattered to gregarious in forests, directly on decaying conifer wood in the fall. Reported from Mt. Angeles on the Olympic Peninsula of Washington State and along the Salmon River near Welches, Oregon. Probably more widely distributed. **Comments:** Activity unknown. *P. washingtonensis* could be weakly psilocybin. It bears similarity to *Psilocybe physaloides* and to a lesser degree *Psilocybe pelliculosa,* although the latter species lacks pleurocystidia. The overall aspect of this mushroom places it into a cluster of difficult-to-delineate taxa, unless microscopic examination is conducted. Several mycenoid Hypholomas (= Naematolomas) look very similar, including *H. udum, H. dispersum,* and allies. See also *Psilocybe silvatica* and *Psilocybe crobula.*

Psilocybe wassoniorum Guzman and Pollock

COMMON NAME: Wasson's mushroom

Cap: 1–2 cm broad. Conic to subcampanulate, sometimes with a slight umbo, smooth, translucent-striate when moist, dark reddish brown to pale brown, fading in drying to an ochre yellow. Flesh thin, fragile, soon bruising bluish where injured. **Gills:** Attachment adnexed, close, pale yellow brown when young, darkening with maturity and becoming dark purplish brown with paler edges. **Stem:** 30–40 mm long by 1–2 mm thick, equal to slightly enlarged, hollow, reddish brown to blackish brown, and covered with a sheath of white floccose fibrillose patches. Narrowing towards the base into a long pseudorhiza up to 10 mm long. Partial veil cortinate, fragile, and soon disappearing. **Microscopic features:** Spores dark purplish brown in deposit, (6) 6.6–7.7 (8.5) by 4.0–5.5 μ, subellipsoid to subrhomboid in side view. Basidia 4-spored. Pleurocystidia absent. Cheilocystidia 14–28 by 5–6.6. μ, ventricose near the base and sublageniform with a flexuous neck 1–2 μ thick, sometimes branched. **Habit, habitat, and distribution:** Solitary or in

***Psilocybe wassoniorum*. Active, potency unknown.**

small groups in open areas of subtropical deciduous forests. Originally found from Veracruz, Mexico, in June and July, at an elevation of 1700–1800 meters. Probably more widely distributed. **Comments:** Active, but potency unknown. Originally found by the late Steven Pollock, and re-collected by Jim Jacobs, this relatively rare species is distinct for its solitary habit and long pseudorhiza. Named to honor the works of the Wassons, this species is not related to *Psilocybe wassonii* (= *Psilocybe muliercula*). See also *Psilocybe herrerae*, a similar species with a long pseudorhiza.

Psilocybe weilii Guzman, Tapia and Stamets

Cap: 2–6 cm broad. Campanulate to bluntly conic with an inrolled margin when young, then incurved, often with an irregular, soon expanding to broadly convex to nearly plane, to uplifted in age. Dark chestnut brown to deep olivaceous brown, typically with a blackish brown zone around the disc, where flesh is 3–4 mm thick at its center. Strongly hygrophanous, fading in drying to pallid brown to light brown. Flesh whitish, bruising bluish. Surface viscid when moist from a separable gelatinous pellicle, translucent-striate near the margin,

Psilocybe weilii nom. prov. Note incurved, convoluted margin, the dark centers on the caps, and the whitish gill edges. Moderately to highly potent.

which can split in age and become tough and opaque in drying. **Gills:** Attachment adnate to sinuate with two tiers of intermediate gills, close, even, broad, light brown overall with pallid, floccose edges. Becoming dark chocolate brown at maturity. **Stem:** 25–70 mm long by 4–8 mm thick. Equal, swelling towards the base, which projects white rhizomorphs. White, becoming dingy brown, bruising bluish overall in age or from drying, covered with a well-developed sheath of whitish fibrillose patches below, and pruinose above. Cartilaginous, strigose, hollow, stuffed with a whitish pith. Partial veil cortinate, leaving a fibrillose annular zone sometimes dusted with purplish violet-brown spores. **Microscopic features:** Spores dark violet grayish black in deposit, subellipsoid in side view, subrhomboid to subellipsoid in face view, 5.5–6.5 by 4–5 µ in side view. Basidia 4-spored. Pleurocystidia abundant, subfusoid or ventricose-rostrate with short apex, or sublageniform, 10.5–21.5 (24) by 5–9.5 (10.5) µ. Cheilocystidia lageniform with short, single, or branched neck, 20–37.5 by 5–6.5 µ. **Habit, habitat, and distribution:** Gregarious to cespitose, sometimes scattered in red-clay soil topped with a thin layer of needles from loblolly pine (*Pinus taeda*) underneath sweetgum (*Liquidambar styraciflua*). First reported from southeastern Cherokee County, in northern Georgia, after hurricane Opal swept through in 1995. Fruiting from early September through November, between temperatures of 45°–80° F, preferring 60°–75°. **Comments:** .61% psilocybin, .27% psilocin, .05% baeocystin, and .32% tryptophan. This is the first report of a lignicolous, caerulescent *Psilocybe* from Georgia. The association with pine needles, along an interface ecosystem (just beyond the edge of an eight-year-old yard lined with shrubs, and in red-clay soils), are habits typical of many other Psilocybes. Additionally, the strong bluing reaction and its tendency to form clusters are characteristics also shared with *Psilocybe caerulescens, Psilocybe baeocystis, Psilocybe aztecorum,* and *Psilocybe heliconia* (Guzman, Tapia, and Stamets 1996). The name honors Andrew Weil and his role in promoting the beneficial properties of mushrooms.

Psilocybe yungensis Singer and Smith

= *Psilocybe acutissima* Heim

COMMON NAME: divinatory mushroom, genius mushroom

Cap: (.5) 1–2 (2.5) cm, conic to campanulate at maturity, often adorned with a sharp umbo. Surface smooth, viscid, and translucent-striate most of the way to the disc when moist, pellicle not separable. Rusty brown to orangish brown to dark reddish brown, hygrophanous, fading in drying to dull yellowish brown or dingy straw colored. Bruising bluish where injured, and then blackish in drying. **Gills:** Attachment adnate to adnexed, close to crowded, dull gray at first, soon purplish brown with spore maturity. Edges pale to nearly concolorous with gill surface. **Stem:** (25) 30–50 (60) mm long by 1.5–2.5 (3) mm thick, equal to enlarging towards the base. Surface covered with a sheath of dense whitish fibrils, pale brownish above and reddish brown to reddish brown–black near the base. Flesh bruising bluish, hollow, and fairly brittle. Partial veil cortinate, soon disappearing with maturity, leaving whitish fibrils along the cap margin and scant remnants on the upper regions of the stem. **Microscopic features:** Spores dark purplish brown in deposit, rhomboid to subrhomboid to subellipsoid, (4.4) 5–6 (7) by 4–6 μ. Basidia 4-spored. Pleurocystidia 14–25 by 4.4–10.5 μ, ventricose below and mucronate at the apex. Cheilocystidia 14–33 by 4.4–7.7 μ, variable, ventricose to clavate to strangulated. **Habit, habitat, and distribution:** Most frequently found in clusters or

Psilocybe yungensis. Probably moderately active.

gregariously on rotting wood, sometimes at the bases of stumps, in coffee plantations or subtropical forests at 1000–2000 meters. Reported from Colombia and Ecuador, and extending north to Mexico, in June and July. Also reported from Bolivia in January. Probably more widely distributed than presently known. **Comments:** Moderately active; analyses not available. This mushroom is distinct for its penchant for growing in great numbers on decomposing stumps or wood debris, its orangish color, and that the caps remain conic at maturity. I find the common name especially revealing. Few species resemble *P. yungensis*. See also *Psilocybe aztecorum*.

Psilocybe zapotecorum Heim emend Guzman

COMMON NAMES: badao zoo, badoo, bei, be-meeche, beya-zoo, bi-neechi, zapos

Cap: 1–3 by 7 (11) cm, highly variable in form, conic to convex to subumbonate, and sometimes papillate and convoluted in age. Surface smooth, translucent-striate near the margin when moist. Reddish brown to organic brown, hygrophanous, fading to beige, orangish rose to straw in drying, quickly bruising blue to green to blackish where injured or in age. **Gills:** Attachment sinuate or adnate, pale brown to violet brown to violet purple, with edges concolorous with gill face, or slightly paler. **Stem:** (40) 100–200 mm long by (3) 5–10 (12) mm thick, equal to slightly expanded at the base, sometimes with a pseudorhiza, at times flexuous or irregular in thickness. White to grayish to variably reddish brown or vinaceous, bluing when touched or injured, with blackish violet tones. Surface floccose above and strongly scabrous-strigose near the base. **Microscopic features:** Spores purplish in deposit, violet brown in deposit, oblong-ellipsoid, (5.5) 6.6–7 (8.8) by 3.8–4.4 (5.5) μ. Basidia 4-spored. Pleurocystidia 20–38 by 5.5–14 μ, variable in form, fusoid-clavate, ventricose to submucronate, sometimes with irregularly divided apices. Cheilocystidia 13–27 by 3.5–6 μ, ventricose to fusoid, pyriform to lageniform, with an extended neck 1.5–2.2 μ thick. **Habit, habitat, and distribution:** Cespitose to gregarious, rarely scattered, in swampy or muddy soils, in humus rich with leaves and wood debris, in marshy deciduous forests, and in coffee plantations. Frequently found on the faces of ravines with exposed soils. Found in southern Mexico (600–1800 meters) and subtropical South America. (Collected in Colombia, Peru, Brazil, and Argentina.) Guzman (1983) reported that this

***Psilocybe zapotecorum.* Moderately to highly active.**

mushroom is sometimes found inside the mud houses of native peoples (Zapotecs), a curious if not spiritually prophetic phenomenon. **Comments:** A potent and strongly bluing mushroom, *P. zapotecorum* is comparatively large and can sometimes be covered with sand as it forces its way up through soils. From Brazilian specimens, Stijve and de Meijer (1993) found up to .30% psilocybin and 1% psilocin, which probably accounts for the strong bluing reaction. Not surprisingly, Heim and Hofmann (1958) found only .05% psilocybin and no psilocin in two-year-old specimens. One of the most curious species in the genus *Psilocybe*, this mushroom has a typically asymmetrical cap that is often convoluted in form. This mushroom is held in high esteem by native Mazatecs and Zapotecs. These two species have been confused frequently. The photographs labeled as *Psilocybe caerulescens* by Ott and Bigwood (1978) and as *Psilocybe caerulescens-zapotecorum* complex by Stamets (1978) are, in fact, *P. zapotecorum*. Heim and Callieux (1959) successfully fruited this species in Erlenmeyer flasks in sterilized, mixed compost after forty days of incubation (24°–26° C).

Psilocybe epilogue

Over the years, many of us have encountered species we could not identify. Here are some photographs of unknown Psilocybes from the files of David Arora, Gary Lincoff, Paul Kroeger, and myself.

Mysterious, unidentified Psilocybes

LEFT: An interesting, active, but unknown *Psilocybe* from a hot spring (grassland/riparian zone) in remote British Columbia, Canada. Similar to *Psilocybe cyanofibrillosa.*

BELOW: An unusual and unknown *Psilocybe*, similar to *Psilocybe semilanceata* except for its large size and unusual cap shape. Found in Washington State.

ABOVE: An unidentified *Psilocybe* from Melbourne, Australia.

RIGHT: An annulate *Psilocybe* from northern Thailand.

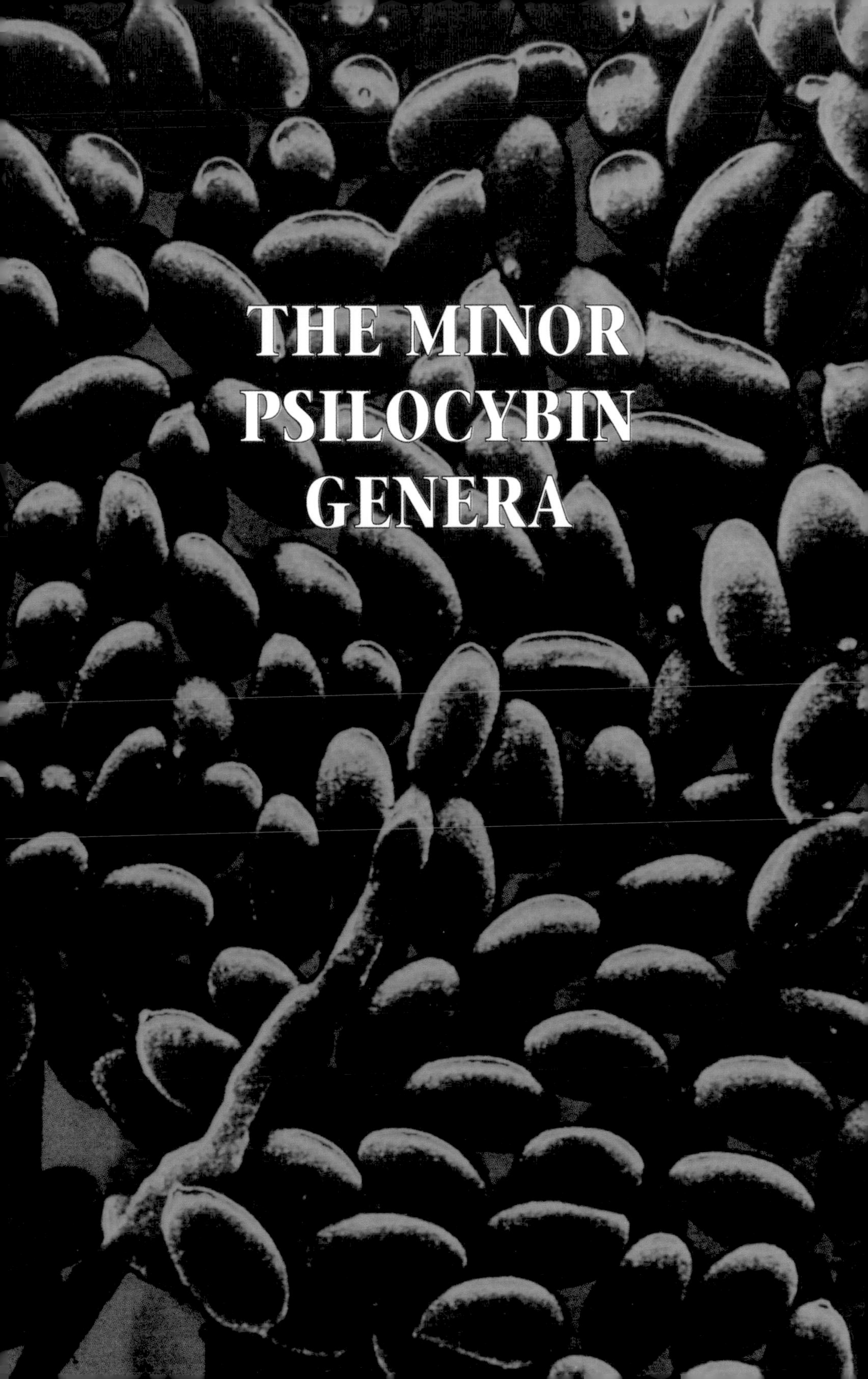
THE MINOR
PSILOCYBIN
GENERA

CHAPTER 9

The Minor Psilocybin Genera

IT SHOULD COME AS NO SURPRISE that tryptamine-derived compounds like psilocybin and psilocin are widespread throughout many genera of mushrooms. Over the past twenty years, numerous research reports have shown psilocybin and psilocin in *Conocybe*, *Gymnopilus*, *Inocybe*, and *Pluteus*. As the biochemical search continues, many more species are likely to be discovered as active. We still have little understanding of what role psilocybin plays in the biology of these species. That so many species produce this unique tryptamine suggests a common biological role. Some state that psilocybin is simply a "waste" product of no significance; others hypothesize that psilocybin is a survival compound and is actively spreading throughout the order of gilled mushrooms (the agaricales). These minor genera pose some risks: there are other members within these genera that are poisonous—a feature not shared with mushrooms within the genera *Psilocybe* and *Panaeolus*. Therefore, please proceed with caution.

The genus *Conocybe*

Conocybes typically grow scattered to gregariously and have a long, thin, fragile stem. The habitats in which this genus grows varies from dung and grass to decayed wood substratum. Species of *Conocybe* that have a well-developed partial veil are placed into the subgenus *Pholiotina,* per Dr. Rolf Singer. *Pholiotina filaris (= Conocybe filaris)* is known to contain toxins similar to those found in the most deadly of Amanitas and Galerinas. It is not very difficult for one to confuse wood- or moss-inhabiting species of *Conocybe* with some species of *Galerina*. A sure way for telling *Galerina* from *Conocybe* is by the microscopic nature of the cap cuticle. Galerinas (and Psilocybes) have filamentous cap cuticles that look like weaved fibers under the microscope, whereas Conocybes (and Panaeoli) have cap cuticles composed of inflated rounded cells resembling cobblestones. To a certain degree, one can visually determine whether a mushroom has a filamentous or cellular cap cuticle by the reflective quality of the cap in moist, fresh specimens. But only the most accomplished experts have developed this skill.

Two species of this genus, *Conocybe cyanopus* and *Conocybe smithii,* are detailed here. Two other species are probably psilocybin-active. *Conocybe siligineoides* was first found in Mexico by Heim and Wasson, and was reported by them to be used for shamanic purposes by the Mazatecs of Oaxaca. Surprisingly, since their discovery circa 1956 no analyses or confirmation of the properties has been published, to my knowledge. A Finnish species, *Conocybe kuehneriana,* has also been reported to be psilocybin active, according to findings by Ohenoja et al. (1987).

With time, both species bruise bluish, although sometimes not markedly so, and usually just at the base of the stem. Both species are very small and lack a ring entirely. But, in view of the existence of the deadly *Pholiotina filaris* and its not-always-present annulus, I strongly caution amateurs who wish to experiment with species of *Conocybe.* There are an abundance of other dark purplish brown–spored Psilocybes and Panaeoli that do not present the dangers inherent within this genus.

Conocybe cyanopus (Atkins) Kuhner

Cap .7–1.2 (2.5) cm broad. Nearly hemispheric to convex, expanding to broadly convex with age. Margin translucent-striate when moist and often appendiculate at first, with minute fibrillose remnants of the partial veil. Reddish cinnamon brown to dark brown. Surface moist when wet, soon dry; smooth overall to slightly wrinkled towards the disc with age. Margin translucent-striate when moist to slightly wrinkled towards the disc. **Gills:** Attachment adnexed, close, and moderately broad. Dull rusty brown with a whitish fringe along the margin. **Stem:** 20–40 mm long by 1–1.4 mm thick. Equal to slightly curved at the base, fragile, easily breaking. Whitish at first, becoming grayish or brownish at the apex, and often adorned with whitish mycelium at the base that bruise bluish. Partial veil thinly cortinate, sometimes leaving trace remnants along the cap margin, soon disappearing. No annulus formed. **Microscopic features:** Spores rusty brown in deposit, 6.5–7.5 (8.5) by 4.5–5 μ. Basidia 4-spored. Pleurocystidia absent. Cheilocystidia 20–25 by 7.5–11 μ. **Habit, habitat, and distribution:** Scattered in grassy areas in lawns

LEFT: *Conocybe cyanopus.* RIGHT: *Conocybe smithii.* Both are moderately to highly active.

and fields in the summer and fall. Reported from Washington, Colorado, Vancouver, B.C., and temperate regions of central and northern Europe (Norway, Finland, and Germany). **Comments:** Potently active, although petite in size. Beug and Bigwood (1982b) found .93% psilocybin but no psilocin. Christiansen et al. (1984) reported ranges of .33–.55% psilocybin and .004–.007% psilocin. Gartz (1992) found .78–1.01% psilocybin, no psilocin, and .12–.20% baeocystin. This species is probably widely distributed across the temperate regions of the world but goes unnoticed because of its minute stature. *Conocybe smithii* is virtually identical except that it favors mossy environments and has longer spores. Beware! Some Galerinas, which can be poisonous, resemble Conocybes.

Conocybe smithii Watling

Cap: .3–1 (1.3) cm broad. Obtusely conic, expanding to nearly plane with a distinct pronounced umbo. Margin translucent-striate most of the way to the disc when moist. Ochraceous tawny to cinnamon brown, hygrophanous, becoming pale pinkish yellow in drying. **Gills:** Adnate to adnexed, soon seceding, crowded to subdistant, narrow to moderately broad. Pale grayish yellow at first, becoming rusty cinnamon brown at maturity. **Stem:** 10–50 (70) mm long by .75–1 (1.5) mm thick. Equal to slightly enlarged at the base, fragile. Whitish, becoming slightly pallid yellowish brown, more grayish at the base. Surface covered with fine fibrils at first but soon smooth overall. Partial veil thin to absent. **Microscopic features:** Spores rusty brown in deposit, (6.5) 7–9 by 4–4.5 (5) μ. Basidia 4-spored. Pleurocystidia absent. Cheilocystidia 20–40 by 9–15 μ. **Habit, habitat, and distribution:** Scattered to numerous in moss in and about sphagnum bogs, and in damp wet places. Reported from Washington, Oregon, and Michigan, probably more widely distributed. Not known from Europe. **Comments:** Probably active, given the bluing reaction, and containing up to .80% baeocystin (Repke et al. 1977). The geographical range of this species is likely to be much more extensive than the literature presently indicates. This species can be found in mossy areas of wet fields. Visually, this species is hard to distinguish from *Conocybe cyanopus*. Microscopically, these two taxa can be separated by spore size.

The genus *Gymnopilus*

The genus *Gymnopilus* has members that give rusty orange to yellow-orange spore prints, are medium to large stature, prefer wood, and typically have dry caps and well-developed veils. Arora (1986) reports there are approximately 75 species of *Gymnopilus* in North America. The number of species worldwide is estimated to be less than 150. To date, 10 species have been shown to be psilocybin-active, according to a survey of the scientific literature by Allen and Gartz (1992). They are *G. aeruginosus, G. braendlei, G. intermedius, G. luteoviridis, G. liquiritiae, G. luteus, G. purpuratus, G. spectabilis, G. validipes,* and *G. viridans* (see also Hatfield et al. 1978). I believe an additional species, *G. luteofolius,* is also active. (The analysis of this species has not yet been reported in the literature.) *G. luteofolius* bruises bluish, especially in cold weather. Additionally, a Mexican *Gymnopilus, Gymnopilus subpurpuratus,* is also likely to be active, given its green bruising reaction. Four of the active *Gymnopilus* species are described here.

All readers should familiarize themselves with the genus *Galerina* before ingesting *Gymnopilus*. Within the genus Galerina, there are deadly poisonous species. Although *Gymnopilus* are generally larger than *Galerina,* both produce rusty spores (rusty orange and rusty brown, respectively) and can have rings. There is a similarity to the untrained eye. A mistake between the two could be deadly. Those who are not skilled at identification should avoid experimenting with *Gymnopilus* species.

Gymnopilus aeruginosus (Peck) Singer

= *Pholiota aeruginosa* Peck

COMMON NAME: magic blue gym

Cap: 2–23 cm broad. Convex with an incurved margin, expanding to broadly convex with age. Variable in color. Dull bluish gray green, to variegated green and yellow, hygrophanous, becoming drab in drying. Surface covered with tawny scales, dark brown in age. Flesh pallid to whitish, with greenish to dull bluish tones, becoming yellowish in drying. **Gills:** Attachment adnexed to adnate to slightly decurrent, often tearing away from the stem, cream buff to pale orangish, close to crowded, broad, with edges often irregular. **Stem:** 30–120 mm long by 4–40 mm thick, similar to cap in color. Surface covered with appressed fibrils, soon disappearing, smooth, slightly striate, solid when young, becoming hollow with age. Partial veil cortinate, yellowish, fragile, soon disappearing and leaving a fibrillose annular zone in the superior region of the stem. **Microscopic features:** Spores rusty brown to rusty orange to reddish cinnamon in deposit, ellipsoid in face view, inequilateral in side view, 6–9 by 3.5–4.5 μ. Basidia 4-spored. Pleurocystidia rare, 23–35 by 5–7 μ. Cheilocystidia 20–38 by 5–9 μ. Flasklike to ventricose, and mostly capitate. **Habit, habitat, and distribution:** Grows gregarious to cespitose on woody debris of hardwoods and conifers, wood chips, sawdust, and stumps from May to September across much of the United States (California, Oregon, Washington, Idaho, Michigan,

***Gymnopilus aeruginosus*. Estimated to be moderately active.**

Tennessee, Ohio, and Pennsylvania), central to northern Europe, and Japan. **Comments:** Moderately active. Few people have actually experimented with this mushroom, probably because it was not widely known to be active. There may be compounds other than psilocybin, but closely related, that potentiate the experiences of the consumer. This large *Gymnopilus* has a bitter taste and is flushed with bluish tones. A slight oily farinaceous odor with hints of anise has been reported from specimens found in the Pacific Northwest.

Gymnopilus luteofolius (Peck) Singer

= *Pholiota luteofolia* (Peck) Saccardo

Cap: 2–6 (8) cm broad. Convex at first, expanding with maturity to broadly convex to nearly plane. Surface dry, covered with dense, often fibrillose, and appressed purple-red to orangish scales, disappearing in age. Dark red to reddish brown, fading in drying to pinkish red or yellowish red, and eventually yellow, sometimes bruising bluish green. Margin even, inrolled to incurved when young, soon straightening, adorned with fibrillose veil remnants. Flesh thick, reddish to purplish, then fading to yellowish in drying. **Gills:** Attachment adnate to sinuate, to slightly decurrent, close to subdistant, broad, initially yellow, then rusty orange, with serrated edges. **Stem:** 30–80 mm long by 3–10 mm thick, equal to enlarged downwards when forming singly, and narrowing when forming in clusters, and often curved at the base. Yellowish or stained rust colored in age. Base of stem sometimes bruising bluish. Partial veil densely cortinate to nearly membranous, leaving a superior membranous annulus or fibrillose annular zone soon dusted with rusty orange spores. **Microscopic features:** Spores rusty orange in deposit, ellipsoid, roughened, 5.5–8.5 by 3.5–4.5 μ. Basidia 4-spored. Pleurocystidia 30–38 by 5–10 μ, fusoid to subventricose. Cheilocystidia 23–28 by 4–7 μ, ventricose to flask shaped with or without a swollen head. **Habit, habitat, and distribution:** Common on woody debris and on wood chips used in nurseries and landscaping, from California to British Columbia from June through November. David Arora has found it along the California coastal region on wood chips of *Eucalyptus* and pine. Hesler (1969) reported that this species is widely scattered across the United States (Texas, Florida, New York, Tennessee, Michigan, and New Mexico) and is primarily found on conifer wood, less commonly on cottonwoods and other hardwoods. David Arora and I

Gymnopilus luteofolius. **Probably weakly to moderately active.**

collected this mushroom on alder (*Alnus rubra*) on Cortes Island in British Columbia. Also occurring on stumps and logs. Fruiting from June through November. **Comments:** Mildly active, according to anecdotal reports. The bluing phenomenon may be temperature driven, with colder temperatures encouraging a stronger reaction. Clones of my collection on Cortes Island produced mushrooms in culture. This mushroom can easily be cultivated on hardwood logs and wood chips using the techniques outlined in *Growing Gourmet and Medicinal Mushrooms*.

Gymnopilus purpuratus (Cooke and Mass.) Singer
= *Flammula purpurata*

Cap: 1–5 cm broad. Cap convex at first with an incurved margin, expanding with age to broadly convex to plane. Reddish purplish brown to purplish red with tinges of yellow and green. Surface covered with scattered fibrillose/floccose patches. **Gills:** Attachment sinuate, waxy yellow, becoming a brownish cinnamon yellow with spore maturity. **Stem:** 20–40 mm long by 2–4 mm thick, stout, narrowing upwards, with fine parallel striations and covered with fine, fragile fibrils. Yellowish brown overall, with greenish and yellowish overtones. **Microscopic features:** Spores bright rusty orange in deposit, 6.5–8.5 by 4.5–5.2 μ, ellipsoid. Lacks cystidia. **Habit, habitat, and distribution:** Grows on rotting wood. In Europe, reported from pig dung mixed with wood chips. Also sighted in south Australia in May. **Comments:** Weakly to moderately active. Gartz (1993) reported that dried, cultivated specimens yielded .23%

psilocybin, .21% psilocin, and .05% baeocystin. Later, Gartz (1994) found roughly similar amounts—.25% psilocybin, .33% psilocin, and .03% baeocystin—also from cultivated specimens. This distinctive mushroom can be cultivated on wood chips or logs using techniques outlined by Gartz (1992). The full range of this *Gymnopilus purpuratus* is not well documented. I would expect it to become more common throughout temperate regions of the world as strains acclimate. Cultivators play a key role in its expansion. *Gymnopilus sapineus*, a closely related species, may also be active but has not yet been analyzed.

Gymnopilus spectabilis (Fries) Singer

COMMON NAMES: big gym, giant laughing mushroom, big laughing mushroom.

Cap: 5–40 cm broad. Convex to broadly convex, expanding to nearly plane with age. Bright yellowish orange, becoming rusty orange to tawny gold to orangish brown or reddish brown at maturity. Surface dry, covered with fibrils or small fibrillose scales. Margin incurved at first, and when young can be decorated with remnants of the veil, straightening or becoming wavy in age. Flesh yellowish. **Gills:** Attachment adnate to sinuate to subdecurrent. Pale yellow to rusty orange, becoming rusty brown with spore maturity. Close to crowded. **Stem:** 30–250 mm long by 1–10 mm thick. Firm, solid, unequal, swelling in the middle or often narrowing towards the base. Rusty orange to yellowish orange, dingy brown towards the base. Surface covered with fine fibrils below the ring. Partial veil densely cortinate to membranous, usually leaving a well-formed, membranous annulus in the superior regions of the stem, soon dusted rusty orange from spores. **Microscopic features:** Spores rusty orange in deposit, ovoid to ellipsoid, roughened, 7–10.5 by 4.5–6 μ. Basidia 2- and 4-spored. Pleurocystidia absent. Cheilocystidia fusoid-ventricose with subcapitate apices, 18–24 by 4–7 μ. **Habit, habitat, and distribution:** Widely distributed across North America, the British Isles, and Europe. In the United States, this mushroom can be found from Texas to California to Washington. Growing gregariously but most commonly in clusters around trees and stumps, occasionally on buried wood. Prefers hardwoods in eastern North America and Europe, and conifers in the western regions of the United States and Canada. **Comments:** Moderately active to inactive. This extremely bitter mushroom varies substantially in its psilocybin content from one geographical region to the next. According to Hatfield et al. (1978), this mushroom is inactive in Europe, but psilocybin-active in Japan, the United States (Ohio, Michigan, Massachusetts), and eastern Canada. Anecdotally, I have heard that races of this mushroom are psilocybin positive in Texas. Analyses of specimens from the western United States proved inactive. Stijve and Kuyper (1988) reported no detectable psilocybin in collections of *G. spectabilis* from Switzerland, the Netherlands, or an unidentified location from the United States. As Beug and Bigwood (1982b) and Smith (1980) noted, many collections identified as *G. spectabilis* are, in fact, *Gymnopilus ventricosus*, a nearly identical but inactive species. (See the photograph on the following page.) In a more recent study of Japanese collections, Tanaka et al. (1993) did not find psilocybin in this species but identified a new hallucinogen, which they described as belonging to a group of "neurotoxic" oligoisoprenoids, with depolarizing activity that was demonstrated on rodent neurons. (No human bioassays were conducted.) There is a pattern in *Gymnopilus* of vast, regional variation in chemical content that needs to be studied further. Three interesting bioassays are worth mentioning. David Arora (1986, 410) reports on the fate of one surprised victim who, after ingesting a *Gymnopilus* and unexpectedly succumbing to its effects, exclaimed on the way to the hospital, "If this is the way you die from mushroom poisoning, then I'm all for it." Andrew Weil and Gary Lincoff, who also

***Gymnopilus spectabilis*, from Japan. Inconsistently active.**

Gymnopilus ventricosus, which is often mistaken for *Gymnopilus spectabilis.* An inactive species.

bioassayed this mushroom (they collected it in Central Park, New York), also experienced classic psilocybin-like symptoms. Finally, there is a remarkable case from 1995 in which a man mistakenly ate a tuft of *G. spectabilis* and experienced not only hallucinations, but also suffered from priapism—for three days! The extremely bitter taste of this mushroom makes it difficult to consume. Given the unusual variability of this mushroom, I urge caution. See also *Gymnopilus validipes.*

Gymnopilus validipes (Peck) Hesler

Cap: 4–30 cm broad. Convex to broadly convex, expanding with age to nearly plane. Bright yellowish orange, becoming rusty orange to tawny gold to orangish brown or reddish brown at maturity. Surface dry, covered with orangish brown fibrils or small scaly patches. Margin deeply incurved at first, and when young can be decorated with threadlike remnants of the partial veil, then decurved and straightening at maturity. Flesh whitish, then yellowish towards the gills. **Gills:** Attachment adnate to uncinate, thick, close to subdistant, yellowish white becoming cinnamon with spore maturity. **Stem:** 100–250 mm long by 25–50 mm thick. Firm, solid, robust, often swelling in the middle, unequal to equal to narrowing towards the solid base. Yellowish white to rusty orange to yellowish orange, usually concolorous with the cap, sometimes dingy brown towards the base. Partial veil cortinate, leaving the trace of an annular zone, and striate at the gill junction. **Microscopic features:** Spores orangish brown in deposit, ellipsoid and slightly ornamented,

7.5–10 by 4.5–5.5 μ, ellipsoid in side view, ovoid in face view. Basidia 4-spored. Pleurocystidia ventricose, inconspicuous, 23–28 by 5–7 μ. Cheilocystidia 23–30 by 3–7 μ. Flask shaped, capitate to subcapitate. **Habit, habitat, and distribution:** Gregarious to cespitose on hardwood logs, stumps, or debris. Widespread in the United States, and undoubtedly distributed across much of North America. Also found in central and northern Europe. **Comments:** Weakly to mildly active, containing 0.12% psilocybin according to Hatfield et al. (1978). (Actual potency may be higher, as psilocin was not screened.) The first reports of activity came from victims who had confused this species with *Armillaria mellea,* the edible honey mushroom (Smith 1980). The mild taste is in contrast to the bitter flavor typical of *Gymnopilus spectabilis,* which shares an affection for hardwoods and looks similar. See also *Gymnopilus ventricosus.*

The genus *Inocybe*

With *Inocybe,* you are truly dancing with danger in a mycological minefield of edible, psychoactive, and toxic mushrooms. Species in the genus *Inocybe* are some of the most difficult of all mushrooms to identify accurately, even for the most experienced mycologists. The late Dr. Daniel Stuntz spent several decades studying this genus, and estimated 400–600 species, of which approximately 150 are recognized. Unfortunately, we know of several poisonous *Inocybe* species, primarily of the muscarinic type. In 1983, when *Inocybe aeruginascens* was mistaken for the common fairy ring mushroom, *Marasmius oreades,* the resulting intoxication led to the analysis that first discovered psilocybin in this genus (Drewitz 1983). Then, Stijve and Kuyper (1985) tested 20 Inocybes with an emphasis on those with greenish gray colors. Five species tested positive for psilocybin. Gurevich and Nezdoiminogo (1994) analyzed 39 *Inocybe* species from Russia, and the only bluing species listed, *Inocybe aeruginascens,* tested positive for psilocybin, while 34 tested positive for the toxin muscarine.

Most Inocybes have not been tested for their edibility, toxicity, or psilocybin activity. Of those species that have been tested, none have *yet* been found to contain both psilocybin and muscarine. There is no reason to believe that the compounds should be mutually exclusive. Will we someday discover one species that is both psilocybin and muscarine producing?

This dearth of information makes the collecting of Inocybes for pleasure a mycological form of Russian roulette. Not only do I stress the utmost caution, but I urge would-be experimenters to defer to the bluing *Psilocybe* or *Panaeolus* species, which are generally more common and much easier to identify. Furthermore, few people have bioassayed the psilocybin-active Inocybes; the reports have come from chemical analyses, not from personal experiences.

Several Inocybes have been found to be psilocybin active. They are *Inocybe aeruginascens, Inocybe corydalina* var. *corydalina, Inocybe corydalina* var. *erinaceomorpha, Inocybe coelestium,* and *Inocybe haemacta* (Stijve and Kuyper (1985). Another *Inocybe, I. calamistrata,* has inconsistently given positive tests for psilocybin (Gartz 1986b). Recently I collected a lone specimen of *I. calamistrata* from an old-growth forest in the Staircase Wilderness region of the Olympic Mountains (Washington) and had it analyzed. The test came back negative, despite the teasing blue-blackening reaction at the base of the stem. I advise that you approach *Inocybe calamistrata* with caution, given these inconclusive analyses. Many Inocybe species, including *I. sororia, I. maculata, I. pudica,* and *I. geophylla,* contain toxic levels of muscarine. With this genus, you are more likely to find one that is toxic before you will find one that is psilocybin active or innocuous.

Inocybe aeruginascens Babos

Cap: 1–5 cm broad. Conic at first, expanding with age to convex and eventually plane with an obtuse umbo. Margin incurved when young, soon straightening. Surface adorned with radial fibrils, more floccose towards the disc. Color sordid buff to sordid ochraceous brown, often with greenish tinges. **Gills:** Attachment adnate to nearly free, crowded, pale grayish brown to clay brown with greenish tones or bruising greenish where injured. **Stem:** 22–50 mm long by 3–7 mm thick. Equal to swelling at the base, solid, whitish to pallid at first, becoming bluish green from the base upwards. Surface pruinose near the apex and longitudinally fibrillose below. Partial veil cortinate, soon disappearing. Flesh whitish, soon bruising bluish green, or naturally bluish green near the base. Odor disagreeable; soapy smelling. **Microscopic features:** Spores clay brown in deposit, smooth, ellipsoid, inequilateral, 7.5–10 by 4–5 μ. Basidia 4-spored. Pleurocystidia 31–71 by 12–24 μ, narrow to broadly fusiform, subclavate, with clear to yellowish tinged walls.

Cheilocystidia scattered, and, when present, similar to pleurocystidia. **Habit, habitat, and distribution:** Widely distributed across the temperate regions of the world. Reported from central Europe and western North America. Found in sandy soils (including dunes), and underneath *Populus* (poplars) and *Salix* (willows) from June through October. **Comments:** Weakly to moderately active, Stijve and Kuyper (1985) reported maxima of 0.28% psilocybin, .008% psilocin, and .08% baeocystin. Gartz (1986b, 1992, 1994) reported .40% psilocybin, no psilocin, .52% baeocystin, and .35% aeruginascin (Gartz's new but uncharacterized indole). Semerdzieva et al. (1986) made more collections of this species and found psilocybin. There are no reports of inactive collections, in contrast to the variability seen in specimens of *Gymnopilus spectabilis*. When found, this mushroom is often covered with sand as it pushes up through riparian soils.

Inocybe coelestium Kuyper

Cap: 1.5–3.2 cm broad. Conic to convex with an incurved margin, expanding with age to broadly convex to nearly plane, and often with a low broad umbo. Ochraceous brown and greenish grayish tinges towards the center. Surface covered with woolly fibrils or covered with uplifted scaly patches. Flesh whitish. **Gills:** Attachment broadly adnate, crowded, narrow, with finely fringed margins. **Stem:** 23–52 mm long by 3–5 mm thick. Equal to swelling at the base, solid, surface smooth or hairy above, and smooth to longitudinally striate below. Whitish near the apex to pale ochraceous, brownish towards the base, often with greenish gray tinges, darkening with age. Flesh brownish with grayish green tones. Partial veil cortinate, soon disappearing. Odor of Peruvian balsam. **Microscopic features:** Spores brown in deposit, smooth, almond shaped, 7–9.5 by 5–6.5 µ. Basidia 4-spored. Pleurocystidia 30–65 by 10–18 µ, cylindrical. Cheilocystidia few, similar to pleurocystidia. **Habit, habitat, and distribution:** Under deciduous trees, beech (*Fagus*) and spruce (*Picea*) in calcareous soils in August through October. Reported from Austria and Germany. **Comments:** Weakly active. Containing up to .035% psilocybin, no psilocin, and .025% baeocystin (Stijve and Kuyper (1985). *Inocybe coelestium* is probably much more widely distributed than reported thus far. This species is relatively rare, and a curiosity in the pantheon of psilocybin fungi, but it is not a good candidate for ingestion by a would-be psychonaut. Be careful.

Inocybe corydalina Quelet

Inocybe corydalina var. *corydalina* Quelet

Inocybe corydalina var. *erinaceomorpha* (Stangl and Veselsky) Kuyper

Cap: 3.8–5.2 cm broad. Obtusely conic to convex at first with an incurved and often denticulate margin, becoming broadly convex to nearly plane in age with or without a broad low umbo. Brown to buff brown, with greenish gray tones, and darker brown to greenish blue, sometimes nearly black, near the center. Surface covered with appressed fibrillose squamules, more densely towards the disc. Flesh white. **Gills:** Attachment narrowly adnate, crowded, broad, and with a minutely fringed margin. Pale brown to buff, or pale grayish brown. **Stem:** 24–95 mm long by 5–15 mm thick. Solid, equal to enlarged near the base. Surface smooth above to fibrillose and longitudinally striate below. Whitish to dull gray below and grayish ochraceous brown to sordid brown overall, with the base often with grayish greenish tinges. Flesh white to gray towards base, slightly reddening upon exposure. Partial veil cortinate, soon disappearing. Scent aromatic, similar to Peruvian balsam. **Microscopic features:** Spores brown in deposit, smooth, lemon to almond shaped, 7–10 by 5–6 μ. Basidia 4-spored. Pleurocystidia 33–70 by 9–21 μ, cylindrical to clavate cylindrical. Cheilocystidia rare and similar to pleurocystidia. **Habit, habitat, and distribution:** Widespread across Europe, the British Isles, and North America in August through October, primarily under deciduous trees (*Fagus*, *Quercus*, *Carpinus*) and to a lesser degree under conifers (*Picea*) in woodland soils. **Comments:** Weakly active, according to Stijve and Kuyper (1985). One variety, *I. corydalina* var. *corydalina*, contains up to .032% psilocybin, no psilocin, and .034% baeocystin. Another, *Inocybe corydalina* var. *erinaceomorpha*, had .10% psilocybin, no psilocin and .034% baeocystin. (Note: the sampling was limited to only three collections.) Gurevich and Nezoiminogo (1994) reported that a collection of *I. corydalina* var. *corydalina* tested negative for psilocybin but positive for muscarine, a result not confirmed by other researchers. *I. corydalina* var. *corydalina* has been reported from both North America and Europe, while *Inocybe corydalina* var. *erinaceomorpha* has thus far only been reported from Europe. The former species has greenish gray fibrils near the disc, while the latter has dark brown scales but without greenish hues. See also *Inocybe coelestium*.

Inocybe haemacta (Berkeley and Cooke) Saccardo

Cap: 1.4–6.5 cm broad. Obtusely conic at first with an incurved margin, soon expanding with maturity to conic convex to broadly convex to plane, and at times uplifted in extreme age with or without an umbo. Greenish gray to dark grayish green towards the center over a pale pinkish brown underlayer, sometimes grayish brown to dark brownish gray near the fibrillose margin and denser towards the center. Surface dry, covered with radially arranged darker fibrils that diverge from the disc, which can be greenish gray to dark gray green. Flesh white to greenish tinged towards the center, becoming pinkish where handled. **Gills:** Attachment narrowly adnate to almost free. Grayish to pale grayish brown, often with olivaceous tones, staining reddish near the edges, becoming reddish brown with spore maturity. **Stem:** 17–85 mm long by 3–10 mm thick. Equal to markedly narrowing towards the base, solid, fine hairs, subfloccose in the upper regions. When young, dingy white to bright pink near the apex and grayish green towards the base, soon darkening to nearly black in age. Flesh pinkish, greenish to dark grayish green at the base, becoming red from exposure to air. Odor is strong, unpleasant, reminiscent of urine. Partial veil cortinate, leaving veil fragile fibrillose remnants on the margin, especially when young. **Microscopic features:** Spores dull clay brown in deposit, smooth, conico-cylindrical, 8–11.5 by 5–6.5 μ. Basidia 4-spored. Pleurocystidia similar to cheilocystidia, sublageniform to lageniform, 55–85 by 14–22 μ. **Habit, habitat, and distribution:** In clay soils in alluvial plains, or in soils enriched with debris, and underneath deciduous woods with oaks (*Quercus*) and beeches (*Fagus*). Widespread throughout Europe but rare in the Netherlands and the British Isles. Probably more widely distributed than presently realized. **Comments:** Weakly active. Containing up to .17% psilocybin, no psilocin, and .034% baeocystin (Stijve and Kuyper 1985). Gartz (1986b) reported psilocybin, psilocin, and baeocystin. This species is named for its distinct colors: the pinkish undertones and pinkish bruising reaction. Although this species is active, I caution readers that since Inocybes are difficult to identify with certainty, and there are poisonous species in this genus, it is wiser to defer to other, easier-to-identify species that are psilocybin active.

The genus *Pluteus*

The genus *Pluteus* features members that are primarily wood decomposers, with caps typically convex to plane, gills pink and free at maturity, and a ringless stem that can be broken away from the cap with ease. Many have pinkish tones in the cap and are fragile, easily breaking and falling apart. Most species are small to midsized and give pinkish to flesh-colored spore deposits. Only one of the active species is illustrated here: *P. salicinus*. Three other *Pluteus* species have shown activity: *P. villosus, P. cyanopus*, and *P. glaucus*. This last species, from Brazil, has up to .28% psilocybin and .12% psilocin, a range comparable to *P. salicinus* (Stijve and de Meijer 1993).

Pluteus salicinus (Persoon ex Fries) Kummer

Cap: 3–7 cm broad. Cap convex to broadly convex, expanding with age to broadly convex to plane. Gray to gray greenish, to bluish gray, darker towards the disc. Surface smooth to finely scaly near the center. **Gills:** Free, not attached. Pallid to cream, soon pinkish to salmon colored at spore maturity. **Stem:** 40–100 mm long by 2–6 mm thick. White to grayish green, often with bluish tones. Flesh often bruising bluish where injured, especially near the base. Base of stem bruising bluish. **Microscopic features:** Spores pinkish in deposit, smooth, ellipsoid to egg shaped, 7–8.5 by 5–6 µ. Pleurocystidia fusiform to lageniform, with or without hooked ends, 58–90 by 10–22 µ and with an apex 5–10 µ thick. Cheilocystidia pear shaped to clavate to cylindrical or slightly lageniform, 30–85 by 8–20 µ. **Habit, habitat, and distribution:** Widely distributed across the United States, the British Isles, and northern Europe. This mushroom is often found in deciduous woodlands in riparian habitats, typically on alder (*Alnus*), willow (*Salix*), or on their woody debris. **Comments:** Weakly to moderately active. Stijve and Kuyper (1985) reported .05–.25% psilocybin, no psilocin, and from zero to .008% baeocystin. Christiansen et al. (1984) found 0.35% psilocybin and .011% psilocin. See also Saupe (1981) and Stijve and Bonnard (1986). The *Field Guide to Mushrooms of Southern Africa* by G.C.A. Van der Westhuizen and Albert Eicker (1994) lists *Pluteus salicinus* as edible although their description lacks any mention of a bluing reaction. This species may have races that vary in their chemical content from region to region, much in the same way as the big laughing mushroom, *Gymnopilus spectabilis*. Caution is advised.

Pluteus salicinus. **Note bluing at base of the stem.**

CHAPTER 10

The Deadly Look-alikes

OF ALL THE POISONOUS MUSHROOMS, the likelihood of an amateur confusing a large, deadly *Amanita* with a small *Psilocybe* is remote. However, in the category of little brown mushrooms (LBMs), several poisonous species could be mistaken for psilocybin varieties if the collector is unaware of basic identification techniques.

Two genera pose the greatest risk to collectors: *Galerina* and *Pholiotina*. Both produce rusty brown spores. The Psilocybes have purplish brown to black spores, and by this distinction alone, you cannot make a deadly mistake between a *Psilocybe* and a *Galerina*. Although the majority of psilocybin mushrooms have dark purplish brown to black spores, there are few psilocybin mushrooms that can be found in the genera *Gymnopilus, Conocybe,* and *Inocybe*. These genera have orangish brown to rusty brown to dull brown spores—the same color of spores shared by some deadly mushrooms. Only skilled mycologists can safely identify the active, psilocybin mushrooms in these genera. Hence, I warn users of this book to avoid these more esoteric species in favor of those that have purplish brown to black spores.

The genus *Galerina*

Galerinas often grow gregariously or in clusters on decaying logs or in moss, and have short brittle stems that often darken from the base. Galerinas may or may not have a partial veil that leaves a membranous ring or fibrillous annular zone. The gills are rusty brown at spore maturity, as is the spore print. This difference in color is critical. Several Galerinas resemble Psilocybes in their overall size and shape. If you cannot distinguish rusty brown from dark purplish brown, your handicap could be life threatening.

The general shape of *G. autumnalis* can be very similar to *Psilocybe stuntzii.* Both are small, squat mushrooms with brown caps and a ring on their stems. Also, there is a great deal of similarity between mycenoid Galerinas, *Psilocybe pelliculosa,* and *Psilocybe semilanceata.* Occasionally, Galerinas can be found in damp, swampy fields, especially near forest edges. Or, woodland Galerinas can pop up through newly laid yards from the rotting wood underneath. So, color is by far the best feature for separating Galerinas from Psilocybes, not shape or habitat.

A mossland mushroom from Idaho, *Psilocybe corneipes* was once thought to be a *Galerina.* This bizarre mushroom has been difficult to place into a genus. First it was an *Agaricus,* then a *Geophila,* then a *Psilocybe,* then a *Galerina.* And now...no kidding...it is in its own genus, *Mythicomyces*! (Who says mycologists don't have a sense of humor?) This bridge species illustrates just how closely Psilocybes and Galerinas can resemble one another. (See Redhead and Smith, 1985.) Watch out for the Galerinas. They can kill you. Most have not yet been analyzed and we may not know of the full range of their toxicity until the first victims fall prey. You don't want to be one of them.

Galerinas can also resemble mushrooms belonging to the genus *Gymnopilus.* Both are orangish brown and produce rusty spores. Because some *Gymnopilus* are psilocybin active, I worry that eager hunters will see the blackening reaction of some Galerinas as a bluing reaction. The mistake of confusing a *Gymnopilus* with a *Galerina* could be deadly. Hence, I discourage the collecting of psilocybin-producing mushrooms that do not have purplish brown to black spores.

At least 199 species of Galerinas are recognized, many of which can only be distinguished microscopically. For this reason, most mycologists, upon sight identification, will place a species such as this into a group of closely related mushrooms such as "*Galerina autumnalis* complex" or simply a "little brown *Galerina.*" Precise identification cannot be made without the use of a microscope. Learn to recognize Galerinas and avoid them.

Galerina autumnalis (Peck) Smith and Singer

COMMON NAMES: autumnal *Galerina,* deadly *Galerina*

Cap: 1–6.5 cm broad. Convex to broadly convex, soon expanding with age to plane, often with an elevated and undulated margin and with a low broad umbo. Margin translucent-striate when moist. Dull yellowish brown to orangish brown to cinnamon or reddish cinnamon

ABOVE: The ring of death: *Galerina autumnalis* fruiting in a rhododendron garden. Conservatively, I would estimate there are enough mushrooms here to kill ten people.

RIGHT AND BELOW: *Galerina autumnalis.*

BOTTOM: *Galerina cinnamomea*, probably deadly.

brown when moist, hygrophanous, fading to dingy yellowish in drying. Surface slightly viscid to viscid when moist from a gelatinous pellicle, sometimes separable. Flesh relatively thin at the margin, and thicker towards the disc. **Gills:** Attachment adnate to sinuate to uncinate, sometimes seceding, close to subdistant, and moderately broad. Golden yellowish brown to rusty brown, becoming dull cinnamon to nearly concolorous with the cap at maturity. **Stem:** 20–60 mm long by (3) 4–8 mm thick. Equal to slightly enlarged at the base and hollow. Grayish brown overall and darkening from the base upwards. Surface covered with fibrillose patches, becoming smooth in age, and often longitudinally striate. Partial veil thinly membranous, leaving a small, fragile membranous annulus that soon deteriorates into a fairly persistent annular zone of fibrils, usually dusted with rusty brown spores at maturity. **Microscopic features:** Spores rusty brown in deposit, ellipsoid, roughened with a lens-shaped depression near the apices, 8–11 by 5–6.5 μ. **Habit, habitat, and distribution:** Grows scattered to gregarious or cespitose in the fall throughout the temperate (mostly western) United States and Canada. Primarily found on decayed conifer and hardwood logs, woody debris, wood chips, bark mulch, or in newly laid lawns. **Comments:** Deadly poisonous! This mushroom was first found to be toxic from people who thought it was *Pholiota mutabilis*, to which it bears a fair resemblance. It contains the same toxins produced by the destroying angels of the genus *Amanita*. Galerinas are similar in their form to *Psilocybe stuntzii*, and are often found growing nearby. Many of the Galerinas have not been tested. Besides *G. autumnalis*, other deadly species are: *G. cinnamomea*, *G. marginata*, and *G. venenata*. The partial veil of *G. marginata* is more cortinate, while that of *G. autumnalis* tends to be more membranous. Many Galerinas can grow in newly placed lawns, in mossy fields, or in soils rich in woody debris. This species has a cap that is more cinnamon red, and may have an annular zone of fibrils on the stem but not a true membranous annulus like *G. autumnalis*. I have collected deadly Galerinas in Colorado, California, Oregon, Washington, and British Columbia. They are widely distributed across much of North America (in the United States as far south as Texas) and across much of Europe.

The genus *Pholiotina*

This is small genus that branched from *Pholiota*. It is listed because of one notorious member: *P. filaris,* a deadly species. *P. filaris* produces the same toxins as the destroying angels (*Amanita virosa, Amanita phalloides, Amanita smithiana,* and allies) and the deadly Galerinas (*G. autumnalis, G. venenata,* and allies). Common in the Pacific Northwest and Europe, *P. filaris* grows from the decorative wood chips used for landscaping around buildings. This species is spreading as a result of landscaping techniques, much in the same manner as the Psilocybes. In the Pacific Northwest, I have found *P. filaris* growing near *Psilocybe baeocystis, Psilocybe cyanescens, Psilocybe pelliculosa,* and *Psilocybe stuntzii.* Furthermore, it is possible that other species closely related to *P. filaris* will become known as poisonous as related species are analyzed. The rusty brown spores and the ring on the stem are warning signs that every collector should learn to recognize.

Pholiotina filaris (Fries) Singer

= *Pholiota filaris* Fries

= *Conocybe filaris* Fries

COMMON NAMES: the deadly *Pholiota,* the deadly ringed cone head, the deadly ringed *Conocybe*

Cap: .5–2.5 cm broad. Obtusely conic at first, soon expanding to conic-convex or convex-campanulate, then nearly plane and often with a pronounced but broad umbo. Margin slightly translucent-striate when moist. Orangish tawny brown. Surface moist when wet, soon dry, usually smooth overall. **Gills:** Attachment adnexed, close, moderately broad, and with one to three tiers of intermediate gills. Rusty brown at maturity. **Stem:** 10–40 mm long by 1–2 mm thick. Stem fibrous, equal to slightly enlarged upwards, and often curved at the base. Dingy yellowish brown to ochraceous. Partial veil membranous, leaving a fragile, nonpersistent, often movable, collarlike membranous annulus, which is rusty brown in color because of the spores in the median to lower regions of the stem. **Microscopic features:** Spores rusty brown in deposit, 7.5–13 by 3.5–5.5 μ. Basidia 2-and 4-spored. Pleurocystidia absent. **Habit, habitat, and distribution:** Scattered to gregarious in decayed wood substratum, in wood or bark chips, or on newly laid lawns and grassy areas that rest upon buried wood. Reported throughout the Pacific Northwest, the British Isles, and Europe. Suspected to be widely distributed. **Comments:** Formerly known as *Conocybe filaris* Fries, this species has been reported to contain deadly toxins similar to *Amanita phalloides* and *Galerina autumnalis*. The variety I frequently find in western Washington has a movable membranous annulus that can degrade into an annular zone, bringing its overall appearance very close to that of other Conocybes. Since there is a potential for misidentification, those not well versed in the identification of mushrooms could conceivably confuse *P. filaris* with psilocybin mushrooms if they did not heed the requirement for a bluing reaction and the other guidelines outlined in this book. Learn to recognize Pholiotinas and Galerinas, and avoid them!

ABOVE and RIGHT: *Pholiotina filaris,* deadly.

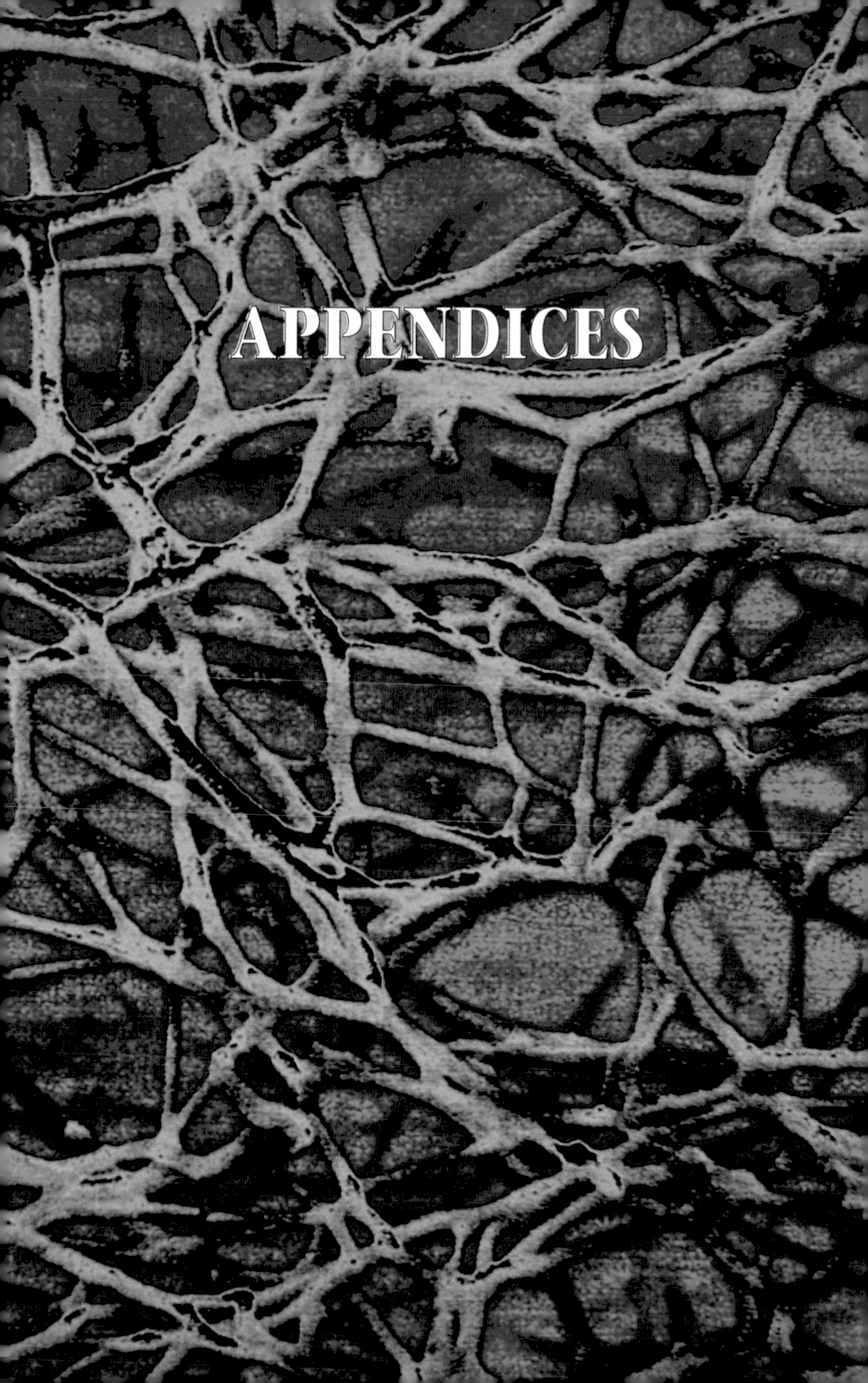
APPENDICES

DIAGRAM A
Cap Shapes

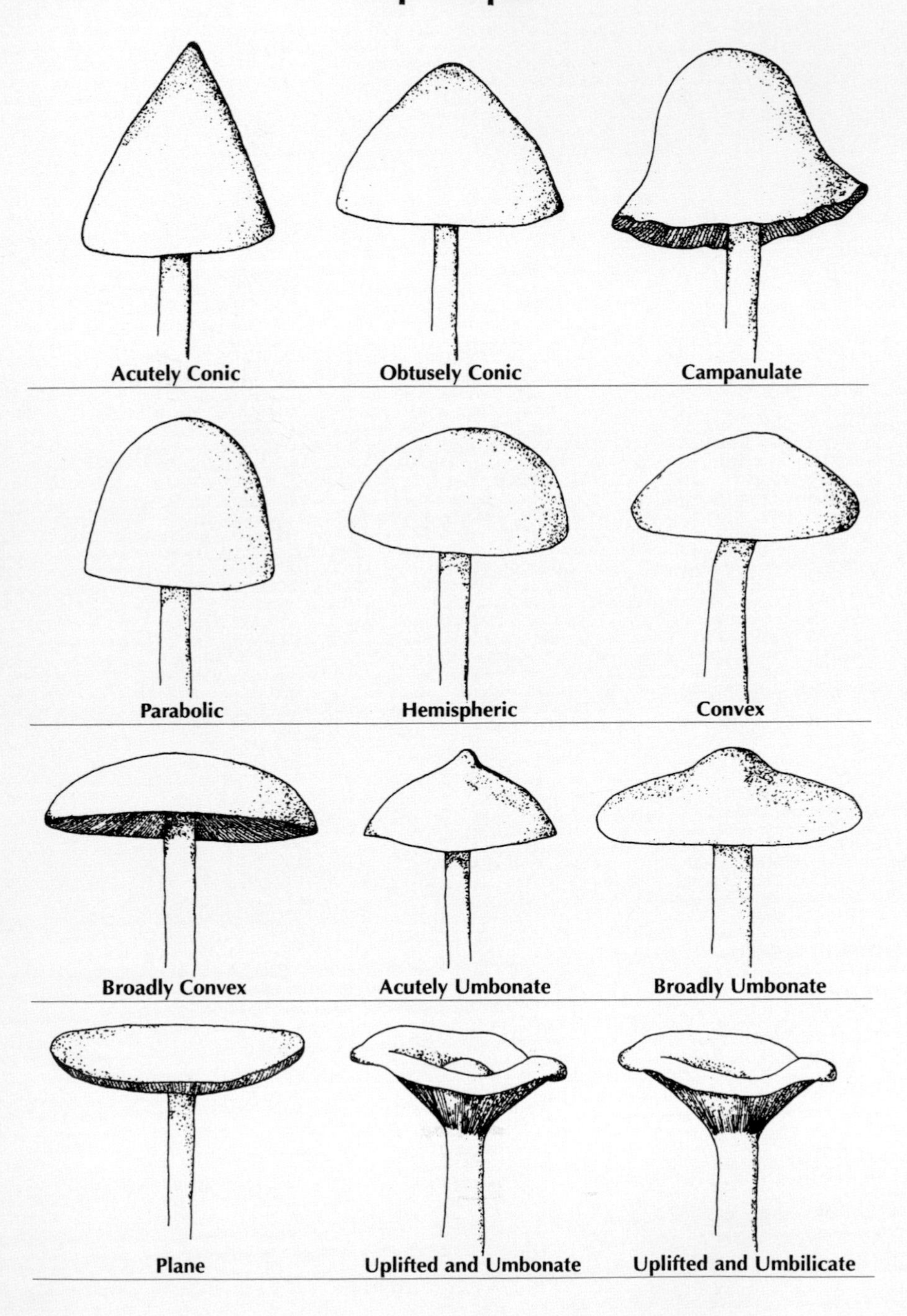

DIAGRAM B
Gill Attachments

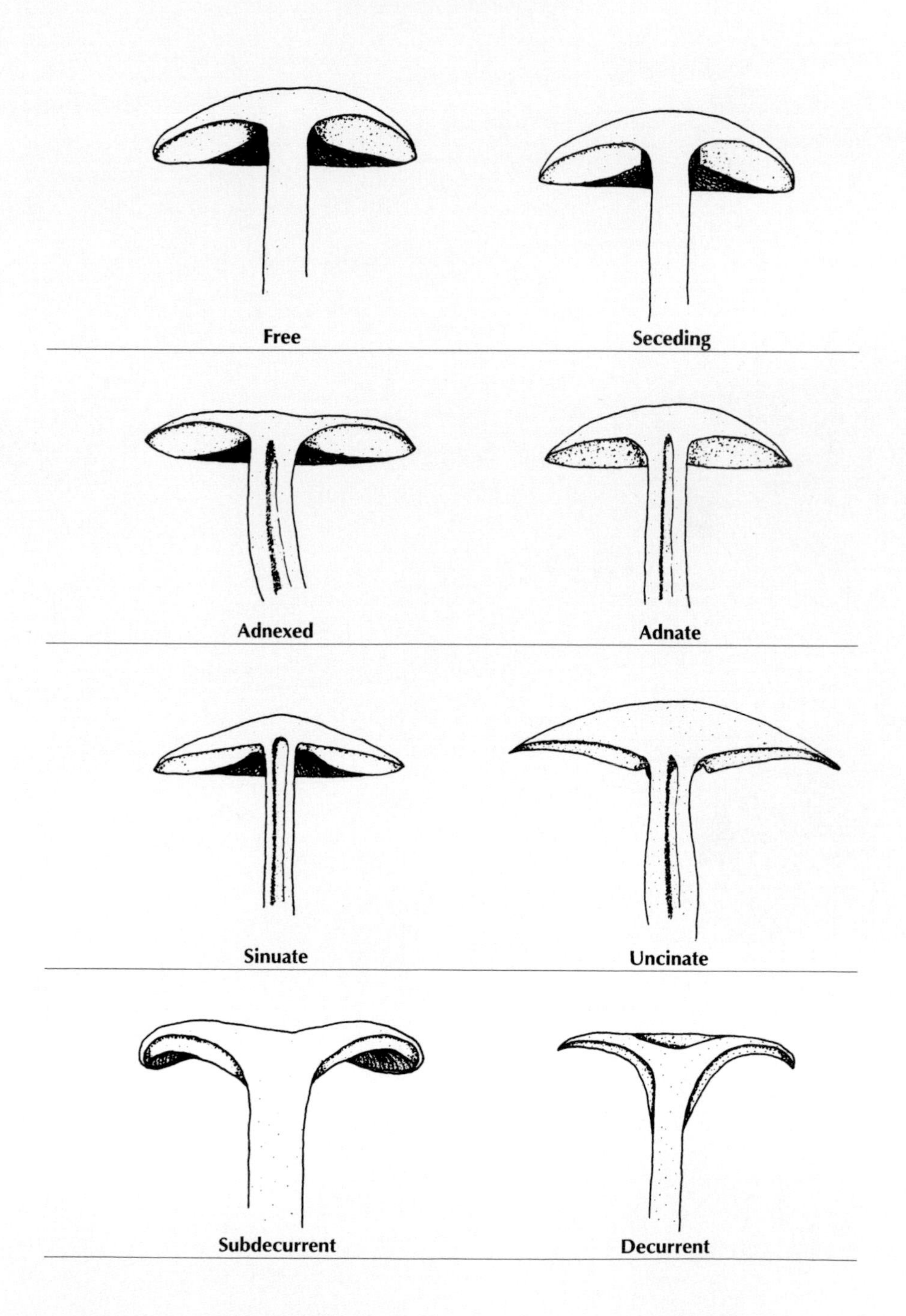

DIAGRAM C

General Morphology

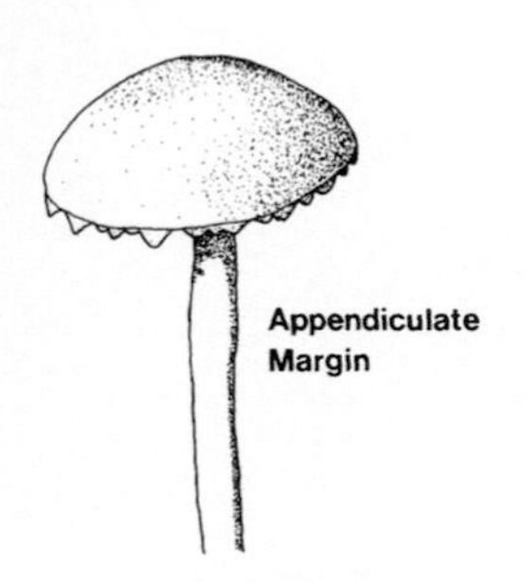

Striated
Margin

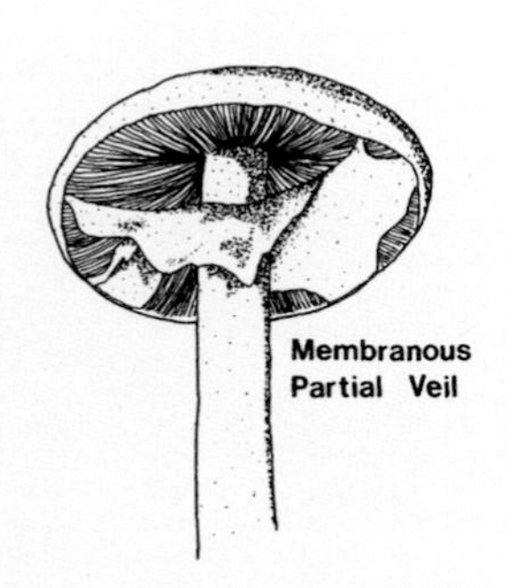

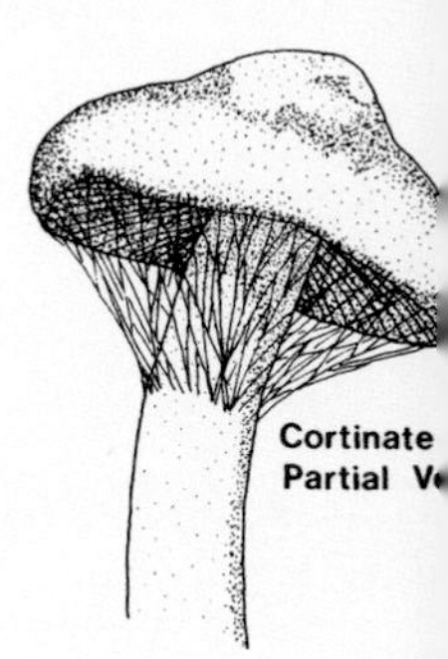

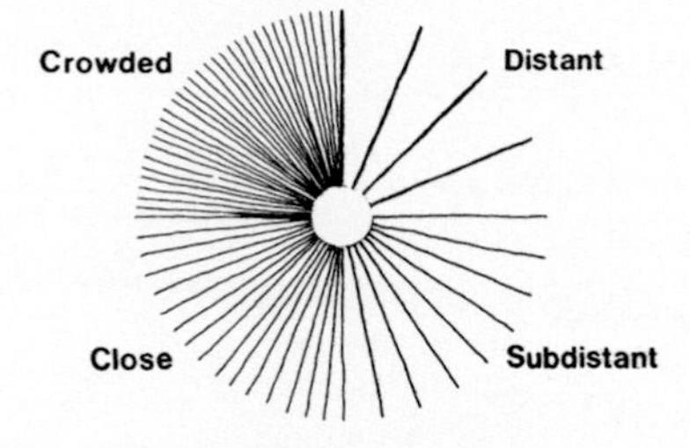

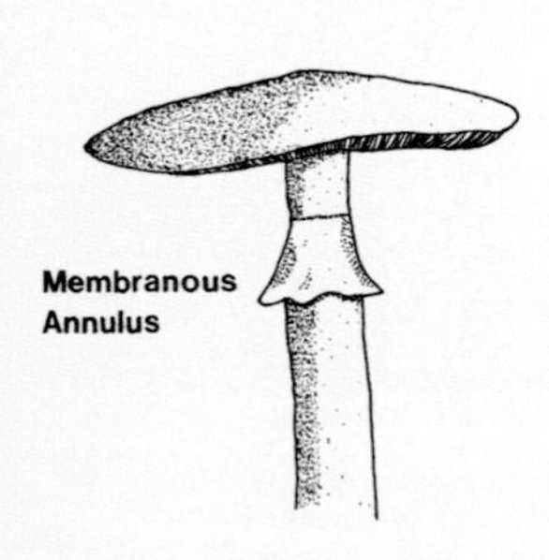

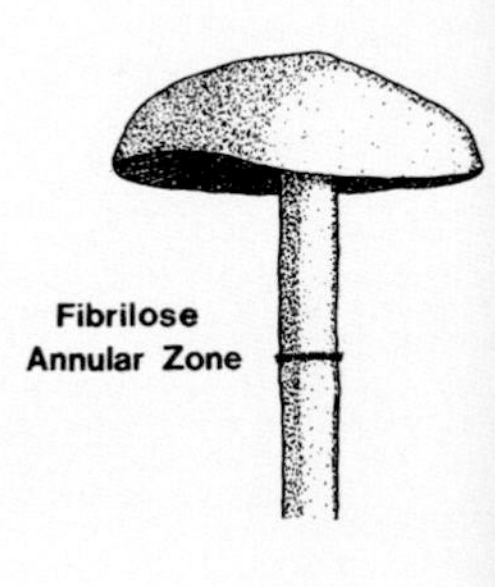

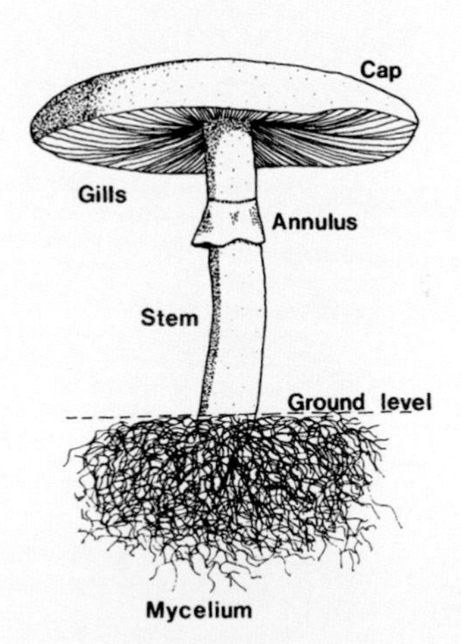

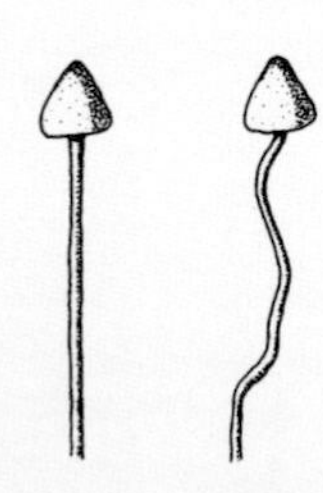

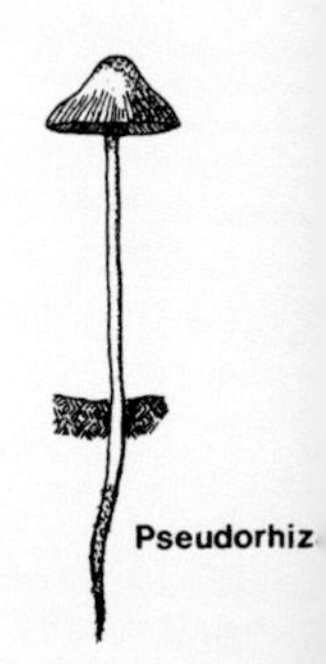

Glossary

acute Pointed, sharp.

adnate (of gills) Bluntly attached to the stem. See Diagram B.

adnexed (of gills) Attached to the stem in an ascending manner. See Diagram B.

Agaricaceae A family of mushrooms with primarily one genus, *Agaricus.*

Agaricales The order that includes all mushrooms with true gills.

agarics Mushrooms with gills.

ampullaceous Flask shaped.

amyloid The characteristic bluish reaction that the flesh or the spores of a mushroom exhibits in Melzer's iodine.

annular (on the stem) Resembling a ring, as in "annular zone." See Diagram C.

annular zone A band of fibrils around the stem, usually becoming darkened by spores. The annular zone is derived from the partial veil whether it be sheathlike (membranous) or cobwebby (cortinate).

annulate Having an annulus.

annulus The tissue remnants of the partial veil adhering to the stem and forming a membranous ring. See Diagram C.

apex (of the stem) The top, or highest point or region.

apiculus The nipplelike projection produced by basidia and to which the spore is attached to the sterigmata of the basidium.

appendiculate (of the cap margin) Hanging with veil remnants.

appressed Flattened.

ascending (of the gills) Where the gills extend upwards from the margin of the cap to their attachment at the stem.

Ascomycetes fungi Produces spores by an ascus, a saclike cell, as opposed to the basidium of Basidiomycetes.

aspect The general shape or outline of a mushroom.

attachment Refers to the attachment between the gills and the stem.

autoclave To steam pressurize.

baeocystin (baeocystine) An indole alkaloid (4-phosphoryloxy-*N*-methyltryptamine), closely related to and often co-occurring with psilocybin. Possibly cerebrally active; comparable to psilocybin.

Basidiomycetes All fungi that bear spores upon a basidium, as opposed to the Ascomycetes, which bear spores in an ascus.

basidia, basidium A particular fertile cell in which meiosis occurs and by which spores are produced.

basidiocarp The fruiting body, the mushroom structure.

binucleate Having two nuclei in one cell.

bioassay A clever term that means that the individual ingested the material to test activity, in effect using one's own body as the diagnostic device.

broad A relative term connoting width (as opposed to length) in reference to the gills (i.e., narrow, moderately broad, broad).

bryophytes Mosses.

buff Dingy yellowish brown.

campanulate (of the cap) Bell shaped. See Diagram B.

cartilaginous Brittle, not pliant.

caulocystidia Sterile cells on the stem.

cellular Composed of globose to generally rounded cells, not thread-like.

cespitose Growing clustered from a common base.

cheilocystidia Sterile cells on the gill edge. Sometimes called marginal cystidia.

chrysocystidia A type of cystidia that is highly refractive in once-dried tissue revived with a potassium hydroxide (KOH) solution. It appears as a yellowish brown amorphous mass within the cell.

clamp connection An elbowlike protuberance that arches over the walls between cells in dikaryotic mycelium of some mushroom species.

clavate: blunted in shape.

close (of the gills) A relative term in reference to the spacing of the gills (in the order crowded, close, subdistant, distant). See Diagram C.

collyboid Resembling a mushroom from the genus *Collybia,* typically with expanded caps (broadly convex to plane), and stems not more than two or three times the breadth of the caps. Often growing in groups or clusters. This term is often used in contrast to mycenoid.

complex A cluster of taxonomically, similarly related species typified by a central species.

concolorous Having the same color.

conic (of the cap) Shaped like a cone.

context A term for the flesh of the cap or stem.

convex (of the cap) See Diagram A.

Copelandian Of or relating to the genus *Copelandia,* distinguished from its close relatives, the Panaeoli, by their dung-loving habitat, strong bluing reaction, and large pleurocystidia.

Coprinaceae A family of mushrooms containing the genera *Coprinus, Panaeolus,* and *Psathyrella.*

coprophilous Growing on dung.

cortina (cortinate) A fine, weblike veil extending from the cap margin to the stem in young specimens of certain species. Soon disappearing or leaving trace remnants on the stem or cap margin.

crowded (of the gills) A relative term used in describing the narrow spacing of the gills (in the order crowded, close, subdistant, distant). See Diagram C.

cuspid (adj. cuspidate) Sharp, pointed.

cuticle The surface layer of cells on the cap that can undergo varying degrees of differentiation.

cystidia Microscopic sterile cells.

decurrent (of the gills) The attachment where the gills markedly run down the stem.

deciduous Describing trees that seasonally shed their leaves.

decurved (of the margin) Where the shape or curvature bends directly downwards.

deliquescing The process of autodigestion by which the gills and cap melt into a liquid. Typical of the *Coprinus* genus and some species of the Bolbitiaceae family.

denticulate Toothlike, lined with triangular fragments of tissue.

dichotomous Repeatedly splitting or forking in pairs.

dikaryophase The phase in which there are two individual nuclei in each cell of the mushroom plant.

dikaryotic The state of cells in the dikaryophase.

diploid A genetic condition wherein each cell has the full set of chromosomes for the species (2N).

disc The central portion of the cap.

eccentric Off-centered.

elevated (of the margin) Describing the type of cap whose margin is uplifted, usually seen in age.

ellipsoid Shaped like an oblong circle.

equal (of the stem) Evenly thick.

eroded (of the margin of the cap or the gills) Irregularly broken.

evanescent Fragile and soon disappearing.

farinaceous Like fresh meal (grain).

fibrillose Having fibrils.

fibrils Fine, delicate hairs found on the surface of the cap or stem.

fibrous (of the stem) Composed of tough, stringlike tissue.

filamentous Composed of hyphae or threadlike cells, which may undergo gelatinization.

flexuous (of the stem) Bent alternately in opposite directions. See Diagram C.

floccose, flocculose Easily removed, usually referring to woolly tufts or cottony veil remnants on the cap or stem.

fruiting body The actual mushroom form or structure; the fruit of the mushroom plant.

fugacious Impermanent, easily torn or destroyed.

fusiform Spindle shaped, tapering at both ends.

fusoid (of the cystidia) Somewhat fusiform.

gelatinous Having the consistency of a jelly, usually translucent.

gill attachment The manner by which the gills meet the stem and/or the cap. **Attached gills** are gills that connect directly to the stem. **Free gills** allow the stem to be separated from the cap without the gills being torn or leaving remnants on the stem.

gills The platelike structures radiating from the underside of the cap and that generate spores.

glabrescent Becoming glabrous.

glabrous Smooth, bald.

gleocystidia A type of pleurocystidia that is highly refractive in tissue that has been revived in potassium hydroxide. Synonymous with chrysocystidia.

glutinous Having a highly viscous gelatinous layer, an extreme condition of viscidity.

gregarious Growing in numerous to dense groups but not clustered, as in "cespitose."

group All the related varieties of one species.

habit The way in which mushrooms are found growing, for example, whether solitary to numerous to cespitose, and the aspect of their forms.

habitat The substrate in which the mushrooms are found.

hemispheric (of the cap) Resembling a hemisphere. See Diagram A.

heteromorphic (of the gill) Used to describe the type of gill edge that is composed of microscopic cell types different from that on the gill surface or face.

homomorphic (of the gill) Used to describe the type of gill edge that is composed of cell types similar to that on the gill surface or face.

humicolous Growing in humus.

hyaline Transparent or translucent, colorless.

hygrophanous Markedly fading in color when drying.

hymenium The layer of fertile, spore-bearing cells on the gill.

hypha, hyphae Individual cells of the mycelium.

incurved (of the margin) Curved inwards.

indigenous Native to a certain region.

KOH Potassium hydroxide; an agent commonly used to revive dried mushroom material for microscopic study at a concentration of 2½% (2½ grams of KOH in 971/2 grams [97.5 milliliters] of distilled water).

lamellae Gills, the spore-producing, platelike structures radially emanating underneath and from the center of the cap.

lamellulae Short intermediate gills not extending the entire distance from the cap margin to the gill attachment.

lanceolate Lancelike.

lignicolous Growing in wood or on a substratum composed of decayed wood.

lignin The basic organic substance of woody tissue other than cellulose.

lubricous Slightly slippery but not viscid.

macroscopic Visible to the naked eye.

mammilate Breastlike, with hardened protuberance, adorned with a nipple.

meiosis The process of reduction division by which a single cell with one nucleus divides into four cells with one nucleus apiece. Each nucleus has one half the genetic material of the parent cell.

membranous Used in describing a homogeneous sheathlike, thin, tissuelike characteristic of a type of partial veil, veil remnant, or some types of volvas.

micron (μ) One millionth of a meter, one thousandth of a millimeter.

mitosis The process of nuclear division in a cell by which the chromosomes are replicated and divided equally to two daughter nuclei.

mononucleate The condition where a cell has only one nucleus.

mottled Spotted, as from the uneven ripening of the spores in the genus *Panaeolus*.

mucronate Pointed, tipped with an abrupt, short point.

mycelium (mycelia) The network of fungi cells that may or may not amass to form the mushroom fruiting body.

mycenoid: Resembling a mushroom from the genus *Mycena:* tall, slender mushrooms with long stems and comparatively small, conic caps. This term is often used in contrast to collyboid.

mycology The science of the study of fungi.

mycophagist A person (or animal) who eats fungi.

mycorrhizal A peculiar type of symbiotic relationship a mushroom mycelium may form with the roots of a seed plant.

naematolomoid (of cystidia) Having the kind of sterile cells on the gill surface that are especially characteristic of the genus *Naematoloma (= Hypholoma)* See chrysocystidia.

naematolomoid (of mushrooms) Resembling species of *Naematoloma (= Hypholoma*)

nanometer One thousandth of a micron.

nomenclature Any system of classification.

norbaeocystin, norbaeocystine 4-phosphoryloxytryptamine, a compound often found in association with baeocystin.

nucleate Having nuclei.

nucleus, nuclei A concentrated mass of differentiated protoplasm in all cells, which plays an integral role in the reproduction and continuation of genetic information to daughter cells.

obtuse Blunt as opposed to pointed.

ochraceous Light orangish brown to pale yellow-brown.

ochre Between warm buff and antimony yellow to ochraceous orange.

olivaceous Olive gray-brown.

ovoid Oval to egg shaped.

pallid Very pale in color, almost a dull whitish.

panaeolian Of or relating to the genus *Panaeolus.*

papilla A small nipplelike protuberance.

partial veil The inner veil of tissue extending from the cap margin to the stem and at first covering the gills in young fruiting bodies of some species. Typically, partial veils are either membranous or cortinate. See Diagram C.

pellicle An upper-surface layer of cells on the cap surface that can undergo gelatinization, making the cap viscid to the touch. Often it can be peeled away from the cap.

persistent Not deteriorating with age; present throughout the life of the fruiting body.

phototropic Attracted or sensitive to light.

pileus The cap of the mushroom.

pileocystidia Sterile cells on the surface of the cap.

pith The central stuffing of stems of some mushrooms.

pleurocystidia Sterile cells on the surface of the gills; sometimes called "facial cystidia."

pliant Flexible.

pore A circular depression evident on the end of spores in many species.

pruinose Having the appearance of being powdered, due to an abundance of caulocystidia on the stem surface.

pyriform Pear shaped.

pseudorhiza A long, rootlike extension of the lower stem.

psilocin, psilocine 4-hydroxy-*N,N*-dimethyltryptamine.

psilocybin, psilocybine *O*-phosphoryl-4-hydroxy-*N,N*-dimethyltryptamine.

psilocyboid Resembling species of *Psilocybe* in general appearance.

psilonaut(s): People who ingest psilocybin mushrooms for spiritual voyages.

radicate With a pseudorhiza, tapering downwards into a narrow root.

reticulate Marked by lines, usually parallel.

rhizomorphs Cordlike strands of twisted hyphae present around the base of the stem.

rhomboid Having an outline resembling a rhombus.

rostrate Fitted with a beak.

scabrous Roughened with short, rigid projections.

scanning electron microscope An electronic microscope that scans an object in a vacuum with a beam of electrons, resulting in a high resolution image that is displayed through a video monitor.

seceding (of the gills) Used to describe the condition where the gills have separated in their attachment to the stem and have the appearance of being free. Often leaving longitudinal lines on the stem where the gills once were connected. See Diagram B.

shamanic A term used to describe any religion that embraces the belief that only special persons or priests are capable of influencing or communicating with the spirits or the supernatural.

sinuate (of the gills) The kind of attachment that seems to be notched before reaching the stem. See Diagram B.

sordid Dingy looking.

spores The reproductive cells or "seeds" of fungi borne on specialized cells.

sporocarps Any mushroom having spores.

squamulose Covered with small scales.

sterigmata Elongated appendages or "arms" on the basidium upon which spores are borne.

stipe Stem of a mushroom.

striate Having radial lines or furrows.

strigose Having long, stiff hairs.

Strophariaceae The family of mushroom containing the closely allied genera *Hypholoma* (*Naematoloma*), *Psilocybe*, and *Stropharia*.

stropharioid Resembling species of *Stropharia*, having a membranous ring, a convex cap, and purplish brown spores.

subclose In reference to the spacing of the gills, between close and crowded. See Diagram C.

subdistant In reference to the spacing of the gills, between close and distant. See Diagram C.

subequal (of the stem) Not quite equal.

substrate, substratum The substance from which mushrooms grow.

superior A term used to designate the location of the annulus in the upper regions of the stem.

tawny Approximately the color of a lion.

taxon (pl. taxa) A taxonomically discrete concept describing a nomenclatural unit, usually a species.

terrestrial Growing on the ground.

trama The fleshy part of the cap beneath the cap cuticle and fertile spore-bearing layer of the gill.

translucent Transmitting light diffusely, semitransparent.

translucent-striate Appearing striate (lined) from the translucent quality of the cap, through which the gills show.

umbilicate (of the cap) Depressed in the center. See Diagram A.

umbo A knoblike protrusion in the center of the cap.

umbonate Having an umbo. See Diagram A.

uncinate A type of gill attachment. See Diagram B.

undulating Wavy. The cap margin of *Psilocybe cyanescens* is a classic example.

universal veil An outer layer of tissue enveloping the cap and stem of some mushrooms, best seen in the youngest stages of development.

variety A subspecies epithet used to describe a consistently appearing variation of a particular mushroom species.

veil A tissue covering the mushrooms as they develop. See definitions of partial veil and universal veil, and Diagram C.

ventricose Swollen or enlarged in the middle.

vinaceous Red-wine colored.

viscid Slimy, slippery when wet or sticky to the touch. In partially dried specimens, it is difficult to tell if the caps were once viscid or not. One test involves touching the cap against one's upper lip. If it sticks, this is a good indication that the cap was once viscid when wet. Typically used in reference to the cap texture. This feature is enhanced under moist conditions.

zonate Having a bandlike zone.

Recommended Reading and Resources

Books

To use this guide to its fullest potential, I strongly encourage you to consult some of the following books to broaden your understanding of mushroom taxonomy. To know only what the psilocybin-producing mushrooms look like, and to have no knowledge of their allies, is dangerous. The more you know, the safer mushroom hunting becomes.

Ainsworth, G.C. *Ainsworth and Bisby's Dictionary of the Fungi*. 8th ed. Kew, England: Commonwealth Mycological Institute, 1995.

Arora, David. *Mushrooms Demystified*. Berkeley: Ten Speed Press, 1986.

———. *All That the Rain Promises and More*. Berkeley: Ten Speed Press, 1991.

Benjamin, Denis R. *Mushrooms: Poisons & Panaceas*. New York: WH Freeman, 1995.

Lincoff, Gary H. *The National Audubon Society Field Guide to North American Mushrooms*. New York: Knopf, 1981.

Snell, Walter, and Esther Dick. *The Glossary of Mycology*. 2d ed. Cambridge: Harvard University Press, 1971.

Stamets, Paul. *Growing Gourmet & Medicinal Mushrooms*. Berkeley: Ten Speed Press, 1993.

Stamets, Paul, and J.S. Chilton. *The Mushroom Cultivator*. Olympia: Agarikon Press, 1983.

Mycological societies

Mycological societies are great resources for identifying mushrooms, as their chartered purpose is primarily educational. They coordinate mushroom forays and mushroom festivals, and often have extensive libraries. Furthermore, mycological societies are networked with one another and with poison-control centers. If you have any questions concerning the identification of a mushroom, feel free to contact them. Be forewarned, however, that some societies will be hesitant to help you identify a psilocybin-containing mushroom. However, they should at least tell you, if they know, whether a mushroom is truly poisonous or only psilocybian.

A national organization known as the North American Mycological Association (NAMA) coordinates activities with the local mushroom societies. This organization can refer you to a local society in your area, if present. NAMA also tracks mushroom poisonings and has set up a network of consulting physicians, versed in all aspects of toxic reactions, should one need to be consulted. NAMA has kept extensive records since 1984 in the form of a "Mushroom Poisoning Case Registry." This registry is made available to Poison Control Centers. The address is

North American Mycological Association
Mushroom Poisoning Case Registry
Blodgett Regional Poison Center
1840 Wealthy S.E.
Grand Rapids, Michigan 49506

Internet resources

General Internet Resource/Forum for Psilocybin Mushrooms:
hyperreal.com/drugs/psychedelics/mushrooms

Terence McKenna's home pages:
personal: http://www.levity.com/eschaton/tm.html
McKennaland: http://www.intac.com/~dimitri/dh/mckenna.html
McKennaArticles: http://www.nepenthes.xo.com/mckenna/index.html

Paul Stamets' home page:
http://www.fungi.com

Visionary Plants/Mushrooms:
ftp://www.nexchi.com/prv/vision/topics/interest/visionary/mushrooms/

(If you know of more Internet sites that should be listed, please e-mail me.)

Works Cited

Abraham, S.P. 1995. Notes on the occurrence of an unusual agaric on *Cymbopogon* in Kashmir Valley. *Nova Hedwigia* 60: 111-117.

Ainsworth, G.C. 1995. *Ainsworth and Bisby's dictionary of the fungi.* 8th ed. Kew, England: Commonwealth Mycological Institute.

Alexopolous, C.H. 1970. *Introductory to mycology.* New York: John Wiley and Sons.

Allen, J.W., and J. Gartz. 1992. Index to the botanical identification and chemical analysis of the known species of the hallucinogenic fungi. *Integration* 2 and 3: 91-97.

Appleton, R.E., J.E. Jan, and P.D. Kroeger. 1988. *Laetiporus sulphureus* causing visual hallucinations and ataxia in a child. *CMAJ* 139: 48-49.

Arora, D. 1986. *Mushrooms demystified.* 2d ed. Berkeley: Ten Speed Press.

Psilocybin and psilocin in certain *Conocybe* and *Psilocybe* species.*Lloydia* 25: 156-159.

Bennedict, R.G., L.R. Brady, and V. E. Tyler. 1962. Occurrence of psilocin in Psilocybe baeocystis. Journal of Pharmaceutical Science. 51: 393–394.

Besl, H. 1994. *Galerina steglichii* spec. nov., ein halluzinogener Haubling. *Z.f. Mykol.* 59 (2): 215-218.

Beug, M., and J. Bigwood. 1981. Quantitative analysis of psilocybin and psilocin in *Psilocybe baeocystis* Singer and Smith by high-performance liquid chromatography and by thin-layer chromatography. *Journal of Chromatography* 207: 370-385.

———. 1982 (a). Variation of psilocybin and psilocin levels with repeated flushes (harvests) of mature sporocarps of *Psilocybe cubensis* (Earle) Singer. *Journal of Ethnopharmacology* 5: 271-291.

———. 1982 (b). Psilocybin and psilocin levels in twenty species from seven genera of wild mushrooms in the Pacific Northwest (U.S.A.). *Journal of Ethnopharmacology* 5: 271-278.

Borhegi, S.F. 1957. Mushroom stones of middle America: a geographically and chronologically arranged distribution chart. In *Mushrooms, Russian and history*.V.P. and R. Gordon Wasson. Appendix to Vol. 2. New York: Pantheon Books.

———. 1961. Miniature mushroom stones from Guatemala. *American Antiquity* 26 (4): 498-504.

Brady, L.R., R.G. Bennedict, V.E. Tyler, D.E. Stuntz, and M.H. Malone. 1975. Identification of *Conocybe filaris* as a toxic basidiomycete. *Lloydia* 38: 172-173.

Chang, Y.S., and A.K. Mills. 1992. Re-examination of *Psilocybe subaeruginosa* and related species with comparative morphology, isozymes and mating compatibility studies. *Mycological Research* 96: 429-441.

Chilton, W.S., and J. Ott. 1976. Toxic metabolites of *Amanita pantherina, A. cothumata, A. muscaria,* and other Amanita species. *Lloydia* 39: 150-157.

Cleland, J.B. 1927. Australian fungi: notes and descriptions. No. 6. *Transactions of the Royal Society of South Australia* 51: 28-306.

Cooke, M.C. 1883. *Handbook of British fungi.* 2d ed. London.

Christiansen, A.L., K.E. Rasmussen, and K. Hoiland. 1981. The content of psilocybin in Norwegian *Psilocybe semilanceata. Planta Medica* 42: 229-235.

———. 1984. Detection of psilocybin and psilocin in Norwegian species of *Pluteus* and *Conocybe. Planta Medica* 45: 341-343.

Dennis, R.W.G., P.D. Orton, and F. B. Hora. 1960. New checklist of British agarics and boleti. *T.B.M.S.* Supp. 43, June.

Drewitz, G. 1983. Ein halluzinogener Risspilizart: Grunlichfarbender Rispliz (*Inocybe aeruginascens*). *Mykol. Mitt.* 26: 11-17.

Earle, F.S. 1906. Algunos (hongos) Cubanos. Inf. An. Estae. *Agron Cuba* 1: 225-242.

Enos, L. 1970. *A key to the American psilocybin mushroom.* Lemon Grove, CA: Youniverse Press.

Furst, P.T. 1972. *Flesh of the gods.* New York: Praeger Publishers.

Gartz, J. 1986a. Quantitative determination of the indole derivatives from *Psilocybe semilanceata* (Fr.) kummer. *Biochem. Physiol. Pflanze* 181: 117-124.

———. 1986b. Untersuchungen zum Vorkommmen des Muscarins in *Inocybe aeruginascens* Babos. *Zeitschrift fur Mikologie* 52 (2): 359-361.

———. 1989. Biotransformation of tryptamine in fruiting mycelia of *Psilocybe cubensis. Planta Medica* 55: 249-250.

———. 1992. New aspects of the occurrence, chemistry and cultivation of European hallucinogenic mushrooms. *Annali Musei Civic di Rovereto* 8: 107-124.

———. 1993. Indole derivatives in certain *Panaeolus* species from East Europe and Siberia. *Mycological Research* 97 (2): 251-254.

———. 1993. *Psychotrope Pilze in Europa.* Editions Heuwinkel, NeuAllschwil/Basel.

———. 1994. Extraction and analysis of indole derivatives from fungal biomass. *Journal of Microbiology* 34: 17-22.

Gartz, J., W. Allen, and M.D. Merlin. 1994. Ethnomycology, biochemistry and cultivation of *Psilocybe samuiensis* Guzman, Bandala,and Allen, a new psychoactive fungus from Koh Samui, Thailand. *Journal of Ethnopharmacology* 43: 73-80.

Gartz, J., and G.K. Muller. 1989. Analyses and cultivation of fruitbodies and mycelia of *Psilocybe bohemica. Biochem. Physiol. Pflanzen* 184: 337-341.

Gartz, J., D.A Reid, M.T. Smith, and A. Eicker. 1996. A new bluing *Psilocybe* from South Africa. *Integration* 6 (in press).

Gerhardt, E. 1987. *Panaeolus cyanescens* (Bk. and Br.) Sacc. and *Panaeolus antillarum* (Fr.) Dennis zwei Adventivarten in Mitteleuropa. *Beitr. Kennin. Pilze Mitteleur* 3: 223-227.

———. 1996. Taxonomische Revision der Gattungen *Panaeolus* und *Panaeolina. Bibliotheca Botanica* 147 (in press).

Goigoux, P. 1992. Un Cas Grave D'Intoxication par *Mycena rosea. Bulletin de la Federation de Mycologie de Dauphine Savoie* 127 (Oct.): 10-11.

Gurevich, L.S. 1993. Indole derivatives in certain *Panaeolus* species from east Europe and Siberia. *Mycological Research* 97 (2): 251–254.

Guzman, G. 1983. *The genus Psilocybe.* Vaduz, Germany: J. Cramer.

———. 1995. Supplement to the monograph of the Genus *Psilocybe.* In *Taxonomic Monographs of Agaricales*. Eds. O. Petrini and E. Horak. *Bibliotheca Mycologica* 159: 91-141.

Guzman, G., V.M. Bandala, and J.W. Allen. 1993. A new bluing *Psilocybe* from Thailand. *Mycotaxon* 46: 155-160.

Guzman, G., V.M. Bandala, and C. King. 1993. Further observations on the genus *Psilocybe* from New Zealand. *Mycotaxon* 46: 161-170.

Guzman, G., and E. Horak. 1978. New species of *Psilocybe* from Papua, New Guinea, New Caledonia, and New Zealand. *Sydowia* 31: 44-54.

Guzman, G., and J. Ott. 1976. Description and chemical analysis of a new species of hallucinogenic *Psilocybe* from the Pacific Northwest. *Mycologia* 68: 1261-1267.

Guzman, G., J. Ott, J. Boydston, and S.H. Pollock. 1976. Psychotropic mycoflora of Washington, Idaho, Oregon, California, and British Columbia. *Mycologia* 68: 1267-1272.

Guzman, G., and A.M. Perez-Patraca. 1972. Las especies conocidas del genero *Panaeolus* en Mexico. *Boletin de la Sociedad Mexicana de Micologia,* no. 6.

Guzman, G., and S. Pollock. 1978. A new bluing species of *Psilocybe* from Florida, U.S.A. *Mycotaxon* 7: 373-376.

———. 1979. Tres nuevas especies y dos nuevos registros de los hongos alucinogenes en Mexico y datos sobre su cultivo en el Laboratorio. *Biol. Soc. Mex. Mic.* 13: 261-270.

Guzman, G., and A. Smith. 1978. Three new species of *Psilocybe* from the Pacific Northwest of North America. *Mycotaxon* 7 (3): 515-520.

Guzman, G., F. Tapia, and P. Stamets. 1997. A new bluing *Psilocybe* from U.S.A. *Mycotaxon* 65: 191–195.

Guzman, G., and R. Watling. 1978. Studies in Australian agarics and boletes. I. Some Species of *Psilocybe. Notes from the Royal Botanic Garden, Edinburgh* 36: 199-210.

Hatfield, G.M., L.J. Valdes, and A. H. Smith 1978. The occurrence of psilocybin in *Gymnopilus* species. *Lloydia* 41: 140-144.

Heim, R. 1963. *Les champignons toxiques et hallucinogenes.* Paris: N. Boubee.

———. 1967. *Nouvelles investigations sur les champignons hallucinogenes.* Paris: Editions du Museum National d'Histoire Naturelle.

———. 1971. A propos des proprites hallucinogenes du *Psilocybe semilanceata. Le Naturaliste Canadien* 98: 415-424.

Heim, R., and R. Caillieux. 1959. Nouvelle contribution a la connaissance des Psilocybes hallucinogenes du Mexique. *Rev. Myc.* 24: 437-441.

Heim, R., and A. Hofmann. 1958. La psilocybine et la psilocine chez les psilocybes et strophaires hallucinogenes. In *Les champignons hallucinogenes du Mexique* 6: 258-267. Paris: Editions du Museum National d'Histoire Naturelle.

Heim, R., A. Hofmann, and H. Tscherter. 1966. Sur une intoxication collective a syndrome psilocybien cause en France par un *Copelandia. Comptes Rendus Hebdomadaire des Seances de l'Academie des Sciences,* Series D, 262: 509-523.

Heim, R., R. G. Wasson, and collaborators. 1958. *Les champignons hallucinogenes du Mexique.* Vol. 6. Paris: Editions du Museum National d'Histoire Naturelle.

Hesler, L.R. 1969. North American species of Gymnopilus. *Mycological Memoirs* 3. New York: Hafner.

Hofmann, A., R. Heim, and H. Tscherter. 1963. Presence de la psilocybine dans une espece european d'agaric, le *Psilocybe semilanceata* (Fr.). *Comptes Rendus Hebdomadaire des Seances de l'Academie des Sciences.*

Høiland, K. 1978. The genus *Psilocybe* in Norway. *Norwegian Journal of Botany* 25: 111-122.

Johnston, P.R., and P.K. Buchanan. 1996. The genus *Psilocybe* (Agaricales) in New Zealand. *New Zealand Journal of Botany* (in press).

Kauffman, C.H. 1971. *The Gilled Mushrooms of Michigan and the Great Lakes Region.* Vol. 1 and 2. New York: Dover Publications.

Keay, S.M., and A.E. Brown. 1990. Colonization of *Psilocybe semilanceata* on roots of grassland flora. *Mycological Research* 94 (1): 49-56.

Klan, J. 1985. *Ceska Mykologie* 39: 58-64.

Koike, Y., K. Yokoyama, K. Wada, G. Kusano, and S. Nozoe. 1981. Isolation of psilocybin from *Psilocybe argentipes* and its determination in specimens of some mushrooms. *Lloydia* 44 (3): 362-365.

Lajoux, J.D. 1961. *Le Meraviglie del Tassili.* Bergamo: Instituto Arti Grafiche.

Lange, J.E. 1923. Studies in the agarics of Denmark. Part 5: Ecological notes. The Hygrophorei. *Stropharia* and *Hypholoma.* Supplementary notes to Parts 1-3. *Dansk Bot. Ark.* 4 (4): 1-55.

Leung, A.Y., and A.G. Paul. 1968. Baeocystin and norbaeocystin: new analogs of psilocybin from *Psilocybe baeocystis. Journal of Pharmacological Science* 57: 1667-1671.

Lhote, H. 1973. A la decouverte des Freques du Tassili. Parigi (Arthaud).

Lhote, H., and K. Nomachi. 1987. Oasis of art in the Sahara. *National Geographic* (August): 181-189.

Liggerstorfer, R., and C. Rätsch, ed. 1996. *Maria Sabina: Botin de Heiligen Pilze.* Nachtschatten (Edition Rauschkunde), Ch-Solothurn, Switzerland.

Lincoff, G. 1995. *The National Audubon Society field guide to North American mushrooms.* New York: Knopf.

Lincoff, G., and D.H. Mitchel. 1977. *Toxic and hallucinogenic mushroom poisoning: a handbook for physicians and mushroom hunters.* New York: Van Nostrand Reinhold Company.

McCawley, E.L., R.E. Brummet, and G.W. Dana. 1962. Convulsions from *Psilocybe* mushroom poisoning. *Proceedings of the Western Pharmacology Society* 5: 27-33.

McKenna, T. 1992a. *The archaic revival.* San Francisco: HarperCollins.

———. 1992b. *Food of the gods.* New York: Bantam.

______. 1993. *True hallucinations.* San Francisco: HarperCollins.

Merlin, M.D., and J.W. Allen. 1993. Species identification and chemical analyses of psychoactive fungi in the Hawaiian islands. *Journal of Ethnopharmacology* 40: 21-40.

Miller, O.K., Jr. 1968. Fungi of the Yukon and Alaska. *Mycologia* 60: 1201-1203.

Miller, O.K., Jr., and D.F. Farr. 1975. *An index of the common fungi of North America (synonymy and common names).* Vaduz, Germany: J. Cramer.

Moser, M. 1983. *Keys to agarics and boleti.* London: Roger Phillips.

———. 1984. Panaeolus alcidis, a new species from Scandinavia and Canada. *Mycologia* 76 (3): 551-554.

Murrill, W.A. 1916. A very dangerous mushroom, *Panaeolus venenosus sp.* nov. *Mycologia* 8: 186-187.

———. 1922a. Dark spored agarics 1. *Drosophila, Hypholoma,* and *Psilocybe. Mycologia* 14: 61-76.

———. 1922b. Dark spored agarics 2. *Gomphidius* and *Stropharia. Mycologia* 14: 121-142.

———. 1923. Dark spored agarics 5. *Psilocybe. Mycologia* 15: 1-55.

Noordeloos, M.E. 1995. Notulae ad floram agaricinam neerlandicam XXIII: *Psilocybe* and *Pholiota. Persoonia* (16): 127-129.

Ohenoja, E., J. Jokiranta, T. Maikenen, A. Kaikkonen, and M. Airaksinen. 1987. The occurrence of psilocybin and psilocin in Finnish fungi. *Lloydia* 50 (4): 741-744.

Ola'h, G.M. 1967. Nouvelle espece de la flore mycologique canadienne. *Naturaliste Canadien* 94: 573-587.

———. 1969a. A taxonomical and physiological study of the genus *Panaeolus* with Latin descriptions of the new species. *Revue de Mycologie* 33 (4): 284-290.

———. 1969b. Le genre *Panaeolus:* essai taxonomique et physiologique. *Revue de Mycologie,* Memoire Hors-Serie 10.

———. 1973. The fine structure of *Psilocybe quebecensis. Mycopathologia et Mycologia Applicata* 49 (4): 321-338.

Ola'h, G.M., and R. Heim. 1967. Une nouvelle espece nord-americaine de *Psilocybe* hallucinogene: *Psilocybe quebecensis* Ola'h and Heim. *Compt. Rend. Acad. Sci.* 264: 1601-1603.

Oss, O.T., and O.N. Oeric. 1986. Psilocybin: magic mushroom grower's guide. Berkeley: Lux Natura.

Ott, J. 1976. *Hallucinogenic plants of North America.* Berkeley: Wingbow Press.

———. 1993. *Pharmacotheon.* Kennewick, WA: Natural Products Co.

Ott, J., and J. Bigwood, ed. 1978. *Teonanacatl: hallucinogenic mushrooms of North America.* Seattle: Madrona Press.

Ott, J., and G. Guzman. 1976. Detection of psilocybin in species of *Psilocybe, Panaeolus*, and *Psathyrella. Lloydia* 39: 258-260.

Ott, J., and S.H. Pollock. 1976. Psychotropic mycoflora of Washington, Idaho, Oregon, California, and British Columbia. *Mycologia* 68: 1267-1272.

Pacioni, G. 1981. *Simon and Schuster's guide to mushrooms.* Gary Lincoff, ed. New York: Simon and Schuster.

Peck, C.H. 1897. *Annual report of the state botanist of the state of New York.* Albany, NY: James B. Lyon.

Phillips, R. 1981. *Mushrooms and other fungi of Great Britain and Europe.* London: Pan Books.

Pollock, S.H. 1975. The psilocybin mushroom pandemic. *Journal of Psychedelic Drugs* 7 (1): 73-84.

———. 1976. Psilocybin mycetisms with special reference to *Panaeolus. Journal of Psychedelic Drugs* 8 (1): 45-56.

Puget Sound Mycological Society. 1972. Mushroom poisoning in the Pacific Northwest. Seattle.

Redhead, S.A. 1985. Book Revue: The genus *Psilocybe. Mycologia* 77: 172-178.

Redhead, S.A. 1989. A biogeographical overview of the Canadian mushroom flora. *Canadian Journal of Botany* 67: 3003-3062.

Redhead, S.A., and A.H. Smith. 1985. Two new genera of agarics based on *Psilocybe corneipes* and *Phaeocollybia perplexa. Canadian Journal of Botany* 64: 643-647.

Repke, D., D. Leslie, and G. Guzman. 1977. Baeocystin in *Psilocybe, Conocybe,* and *Panaeolus. Lloydia* 40: 566-578.

Robbers, J.E., V.E. Tyler, and G.M. Ola'h. 1969. Additional evidence supporting the occurrence of psilocybin in *Panaeolus foenisecii. Lloydia* 32: 399-400.

Rumack, B.H., and D.G. Spoerke. 1995. *Mushroom poisoning: diagnosis and treatment.* West Palm Beach, FL: CRC Press.

Sahagun, B. 1985. *The Florentine codex, general history of the things of New Spain.* Santa Fe, NM: The School of American Research.

Samorini, G. 1995. The oldest representations of hallucinogenic mushrooms in the world. *Integration* 2 (3): 69-68.

———. *Gli allucinogeni nel mito: racconti sull'origine delle piante psiocative.* Torino, Italy: Nautilus.

Samorini, G., and G. Camilla. 1995a. Rappresentazioni fungine nell'arte greca.*Ann. Mus. Civ. Rovereto.*

———. 1995b. Uso trasizionale di funghi psicottaivi in Costa d'Avorio. (Traditional use of psychoactive mushrooms in the Ivory Coast.) *Eleusis* 1: 22-27.

Saupe, S.G. 1981. Occurrence of psilocybin/psilocin in *Pluteus salicinus* (Pluteaceae). *Mycologia* 73: 781.

Schultes, R.E. 1939. The identification of teonanacatl, a narcotic basidiomycetes of the Aztecs. *Plantae Mexicanae II, Botanical Museum Leaflets,* Harvard University 1 (3): 37.

Sebek, S. 1980. Bohmischer Kahlkopf, *Psilocybe bohemica. Ceska Mykologie* 37: 177-181.

Semerdzieva, M., T. Koza, and J. Gartz. 1986. Psilocybin in Fruchtkorpern von *Inocybe aeruginascens. Planta Med.* 47: 83-85.

Singer, R., and A. H. Smith. 1958a. New species of *Psilocybe. Mycologia* 50: 141-142.

———. 1958b. Mycological investigations on *teonanacatl* the Mexican hallucinogenic mushroom. Part 1: the history of *teonanacatl,* field work and culture work. Part ll: a taxonomic monograph of *Psilocybe,* section Caerulescentes. *Mycologia* 50: 239-303.

Singer, R. 1975. *The agaricales in modern taxonomy.* 3d ed. Vaduz, Germany: J. Cramer.

Smith, A.H. 1946. New and unusual dark spored agarics from North America. *Journal of the Mitchell Society* 4: 195-200.

———. 1947. *North American species of mycena.* Ann Arbor: University of Michigan Press.

———. 1948. Studies on dark-spored agarics. *Mycologia* 40: 669-707.

———. 1949. *Mushrooms in their natural habitats.* New York: Hafner Press.

———. 1951. North American species of *Naematoloma. Mycologia* 43: 468-515.

———. 1979. Generic relationships within the Strophariaceae of the agaricales. *Taxon* 28: 19-21.

———. 1980. *The mushroom hunter's field guide.* Ann Arbor: University of Michigan Press.

Smith, A.H., and R. Singer. 1964. *A monograph on the genus Galerina Earle.* Portland, OR: Sawyer's.

Snell, W.H., and E.A. 1971. *A glossary of mycology.* Cambridge: Harvard Univ. Press.

Stamets, P. 1995. *Growing gourmet and medicinal mushrooms.* Berkeley: Ten Speed Press.

———. 1978. *Psilocybe mushrooms and their allies.* Seattle: Homestead Book Co.

Stamets, P., M. Beug, J. Bigwood, and G. Guzman. 1980. A new species and a new variety of *Psilocybe* from North America. *Mycotaxon* 11: 476-484.

Stamets, P., and J.S. Chilton. 1983. *The mushroom cultivator.* Olympia: Agarikon Press.

Stamets, P., and J. Gartz. 1995. A new caerulescent *Psilocybe* from the Pacific Coast of northwestern North America. *Integration* 6: 21-27.

Stijve, T.C. 1987. Vorkommen von Serotonin, Psilocybin und Harnstoff in Panaeoloideae. *Bertrage zur Kenntnus der Pilze Mitteleuropas 3*: 229-234.

———. 1988. Absence of psilocybin in species of fungi previously reported to contain psilocybin and related tryptamine derivatives. *Persoonia* 13 (4): 463-465.

———. 1992. Psilocin, psilocybin, serotonin, and urea in *Panaeolus cyanescens* from various origins. *Persoonia* 15 (1): 117-121.

———. 1994. Bioconcentration of manganese and iron in Panaeoloideae Sing. *Persoonia* 15 (4): 525-529.

———. 1995. Worldwide occurrence of psychoactive mushrooms: an update. *Czech Mycol.* 48 (1): 11-19.

Stijve, T.C., and J. Bonnard. 1986. Psilocybine et urée dans le genre *Pluteus. Mycologia Helvetica* 2: 123-130.

Stijve, T.C., and A.A.R. de Meijer. 1993. Macromycetes from the state of Parana, Brazil. 4. The psychoactive species. *Arq. Biol. Technol.* 36 (2): 313-329.

Stijve, T., C. Hischenhuber, and D. Ashley. 1984. Occurrence of 5- hydroxylated indole derivatives in *Panaeolina foenisecii* (Fries) Kuehner from various origins. *Zeitschrift fur Mykologie* 50: 361-368.

Stijve, T.C., J. Klan, and T.W. Kuyper. 1985. Occurrence of psilocybin and baeocystin in the genus *Inocybe* (Fr.) Fr. *Persoonia* 12 (4): 469- 473.

Stijve, T. C., and T.W. Kuyper. 1985. Occurrence of psilocybin in various higher fungi from several European countries. *Planta Medica* 51 (5): 385-387.

Tanaka, M., K. Hashimoto, T. Okuno, and H. Shirahama. 1993. Neurotoxic oligoisoprenoids of the hallucinogenic mushroom *Gymnopilus spectabilis*. *Phytochemistry* 34: 661-664.

Van der Westhuizen, G. C.A., and A. Eicker. 1994. *Field guide to mushrooms of South Africa*. Cape Town: Struick Publishers.

Verrill, A.E. 1914. Discussion and correspondence. *Science* 40: 408-410.

Wakefield. 1946. *Transactions of the British Mycological Society* 29: 141.

Wasson, R.G. 1958. The divine mushroom: primitive religion and hallucinatory agents. *Proceedings of the American Philosophical Society* 102: 221-223.

———. 1961. The hallucinogenic fungi of Mexico: an inquiry into the origin of the religious idea among primitive peoples. *Harvard Univ. Botanical Museum Leaflets* 19: 147.

———. 1972. *Soma, divine mushroom of immortality*. New York: Harcourt Brace Jovanovich.

Wasson, R.G., F. Cowan, and W. Rhodes. 1974. *Maria Sabina and her Mazatec mushroom velada*. New York: Harcourt Brace Jovanovich.

Wasson, R.G., and N.M. Gregory. 1987. *British fungus flora: agarics and boleti*. 5. *Strophariaceae and Coprinaceae Hypholoma, Melanotus, Psilocybe, Stropharia, Lacrymaria and Panaeolus*. 121. Edinburg: Royal Botanic Gardens.

Wasson, R. G., and R. Heim. 1959. The hallucinogenic mushrooms of Mexico: an adventure in ethnomycological exploration. *Transcripts of the New York Academy of Sciences* 21: 325-339.

Wasson, R.G., A. Hofmann, and C.A.P. Ruck. 1978. *The road to Eleusis: unveiling the secret of the mysteries. Ethno-mycological studies 4*. New York: Harcourt Brace Jovanovich.

Wasson, V.P., and R.G. Wasson. 1957. *Mushrooms, Russia, and history*. New York: Pantheon Books.

Watling, R., and N.M. Gregory. 1987. Stophariaceae and Coprinaceae *Hypholoma, Melanotus, Psilocybe, Stropharia, Lacrymaria, and Panaeolus*. British Fungus Flora: Agarics and Boleti. 5. Edinburg: Royal Botanic Gardens.

Weeks, R.A., R. Singer, and W.L. Hearns. 1979. A new species of *Copelandia*. *Journal of Natural Products* (*Lloydia*) 42 (5): 469-474.

———. 1975. Mushroom hunting in Oregon. *Journal of Psychedelic Drugs* 7: 89-102.

———. 1977. The use of psychoactive mushrooms in the Pacific Northwest: an ethnopharmacological report. *Botanical Museum Leaflets of Harvard* 25 (5): 131-148.

Weil, A. 1986. *The natural mind: a new way of looking at drugs and the higher consciousness*. New York: Houghton Mifflin.

Weil, A., and W. Rosen. 1993. *From chocolate to morphine: everything you need to know about mind-altering drugs*. New York: Houghton Mifflin.

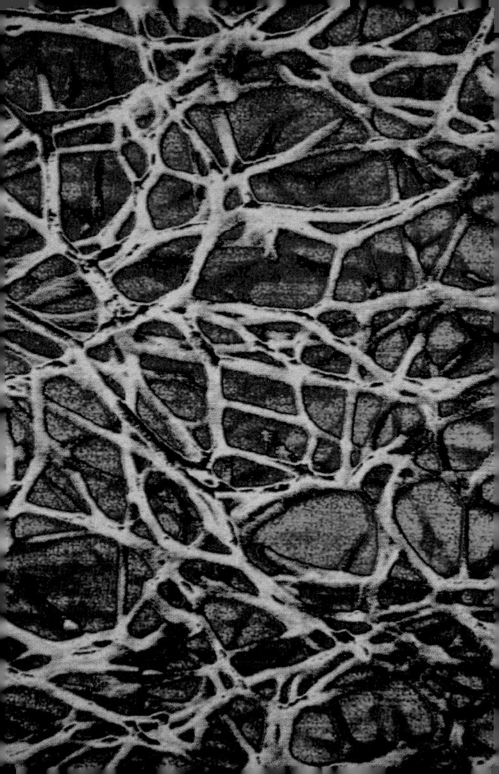

Acknowledgments

FOREMOST, I THANK THE PSILOCYBES, who have been my greatest teachers.

Many people played critical roles in the creation of this book. If it were not for the works of Gaston Guzman, Jim Jacobs, Jochen Gartz, Gary Lincoff, and David Arora, this book would not have been possible. My immediate family has provided many hours of able assistance, tolerating my incessant preoccupation with writing. Thanks, Azureus, and LaDena.

Those participating in the production of this treatise who deserve my thanks are Tammy Davis, Hal Hershey, Catherine Jacobes, Donna Latte, Andrew Lenzer, and Kirsty Melville. Others who deserve acknowledgment who have helped in their own ways are John W. Allen, Joseph Ammirati, Alan and Arleen Bessette, Paxton Hoag, Eric Iseman, Paul Kroeger, Terence McKenna, Gary Menser, Meinhard Moser, Jonathan Ott, Christian Ratsch, Scott Redhead, Giorgio Samorini, Catherine Barnhart-Scates, Alexander Smith, T. Stijve, Roy Watling, Andrew Weil, and Phil Wood. Steve Rooke is thanked for his color enhancement of my scanning electron micrographs. To all of you, I am grateful for your encouragement and contribution.

Last thanks go to the Evergreen State College for sponsoring much of my earlier work with *Psilocybe,* to the BOTS group for the inspiration to continue on this path, and to the Botanical Preservation Corps, who made my trips to Mexico possible and interesting.

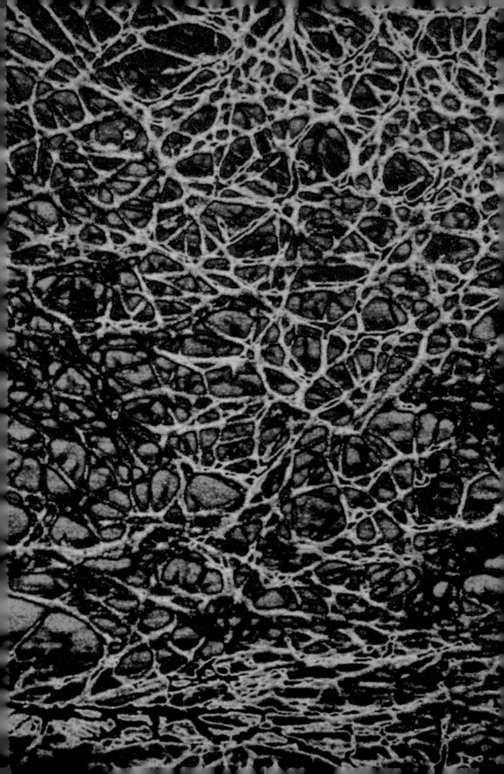

Index

C

F

G

H

N

O

Q

R

S

T

About the Author

PAUL STAMETS HAS BEEN STUDYING MUSHROOMS for more than twenty years, and has discovered and coauthored four new psilocybin mushrooms: *P. azurescens, P. cyanofibrillosa, P. liniformans* var. *americana,* and *P. weilii.*

He runs a mushroom mail-order business, Fungi Perfecti, which grows and ships gourmet and medicinal (no psilocybin) mushroom products, and conducts in-depth workshops on mushroom cultivation. (For a catalog, call 360/426-9292, fax 360/426-9377, or e-mail at mycomedia@aol.com. Paul's home page is www.fungi.com.)

His previous books include *Psilocybe Mushrooms & Their Allies, The Mushroom Cultivator,* and *Growing Gourmet & Medicinal Mushrooms.*